W9-BRD-438

skin

skin

A NATURAL HISTORY

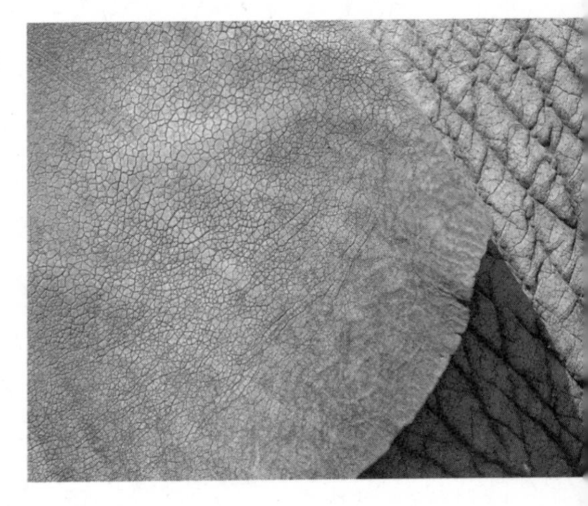

NINA G. JABLONSKI

UNIVERSITY OF CALIFORNIA PRESS *Berkeley Los Angeles London*

University of California Press, one of the most distinguished university presses in the United States, enriches lives around the world by advancing scholarship in the humanities, social sciences, and natural sciences. Its activities are supported by the UC Press Foundation and by philanthropic contributions from individuals and institutions. For more information, visit www.ucpress.edu.

University of California Press
Berkeley and Los Angeles, California

University of California Press, Ltd.
London, England

© 2006 by Nina G. Jablonski

Title page photograph © Heather Leah Kennedy

Library of Congress Cataloging-in-Publication Data

Jablonski, Nina G.
Skin : a natural history / Nina G. Jablonski.
p. cm.
Includes bibliographical references and index.
ISBN-13: 978-0-520-24281-4 (cloth : alk. paper),
ISBN-10: 0-520-24281-5 (cloth : alk. paper)
1. Skin. I. Title.

QP88.5.J33 2006
612.7'9—dc22 2006007731

Manufactured in Canada

15 14 13 12 11 10 09 08 07
10 9 8 7 6 5 4 3

This book is printed on New Leaf EcoBook 50, a 100% recycled fiber of which 50% is de-inked post-consumer waste, processed chlorine-free. EcoBook 50 is acid-free and meets the minimum requirements of ANSI/ASTM D5634–01 (*Permanence of Paper*).

TO GEORGE

CONTENTS

ILLUSTRATIONS

Color Plates and Maps

Following page 82

Plates

Maps

Figures

ACKNOWLEDGMENTS

Books arise in the heart and then grow and take shape after a long gestation, nourished by conversations, research, chance encounters, contemplation, and simple toil. I first thought, many years ago, that I would write a book on skin color because I knew something about the subject and recognized its importance. After talking over the prospect of such a project with Blake Edgar, my editor at the University of California Press, I realized that a book with a wider scope—the story of skin itself, not simply skin color—was clearly needed. I knew of no other work that dealt with human skin in its entirety, in the manner of an old-fashioned natural history. I have Blake to thank for many things, but foremost among them is for enticing me to broaden the scope of my original idea, thus allowing me to deepen and enrich my knowledge about one of the most important parts of the body.

The writing of this book took me to a wondrous and lurid array of places, on the ground and in cyberspace. Some of them, such as anthropology museums and human anatomy laboratories, were already very familiar to me; others, such as tattoo conventions and the Web sites of performance artists and manufacturers of implantable microchips, were refreshingly new. I had a great respect for skin when I started this work and a substantially greater measure of respect by the time I finished. Skin is not only crucial to our health; it is also an important vehicle for self-expression. The eyes may be the mirror of the soul, but the skin is the mirror of everything else.

I was able to write this book because of the support and active assistance of many people and organizations. Thanks are owed to four people in particular, without whose individual input the book would never have been completed. Blake Edgar acted as a muse and an unfaltering guide throughout the life of this project. Good editors do many things for their authors, serving alternately as psychotherapist, grammarian, content specialist, creative writing teacher, and life coach, as needed. Blake filled these varied roles effectively and effortlessly and never made me feel like an idiot even when at times I wrote and acted like one. He also enlisted four reviewers whose varied, thoughtful, and constructive comments made this book more readable, comprehensive, and approachable.

Nana Naisbitt, my friend and supporter, offered both inspiration and assistance. In her capacity as the executive director of the Pinhead Institute in Telluride, Colorado, she and her board chose me as the first Pinhead Institute Scholar in Residence in July 2005. I am grateful that I was able to write most of the first draft of my manuscript in such a beautiful place.

Bonnie Warren, my research assistant, helped me assemble and track most of the references and illustrations that appear in the book. This was a Herculean task, which required patience, perseverance, and punctilious attention to detail. Her unflappable attitude and good humor were invaluable.

George Chaplin, my husband, read the first draft of the book and offered many insightful suggestions and constructive criticisms that improved it. Conversations with him over the years have been my greatest single source of inspiration. George also provided me with just the right diet of advice, good food, cajoling, jokes, and badgering that I needed to finally get the job done. I doubt that this book could have been written without his love and enduring, steadfast support.

I received generous advice and access to useful photographic images from many colleagues and friends in the course of researching and writing. These included Robert Altman, Mauricio Antón, Victoria Bradshaw, Alastair Carruthers, James Cleaver, Paul Ekman, Harriet Fields, David Kavanaugh, Patrick V. Kirch, Dong Lin, Kira Od, Edward S. Ross, Geerat Vermeij, and Christopher Zachary. From the library of the California Academy of Sciences, Patti Shea-Diner helped me with innumerable interlibrary loans; and Kathleen Berge and Anna Barr, of the Anthropology and Education Departments, respectively, provided generous assistance in the final stages of manuscript preparation. The excellent editorial and production staff at the University of California Press worked tirelessly to make this a readable and attractive book. My special thanks go to Dore Brown, Matthew Winfield, Nicole Hayward, and especially Mary Renaud for her superb copyediting.

Finally, I would like to thank the Fletcher Foundation for providing me with an Alphonse Fletcher Sr. Fellowship during the final months of writing this book. This financial support allowed me to include many more illustrations in the book than otherwise would have been possible. Thanks to that generous award and the research I have done in the last few years, the book on skin color that started me off on this project several years ago has been born anew in my heart.

INTRODUCTION

Our skin mediates the most important transactions of our lives. Skin is key to our biology, our sensory experiences, our information gathering, and our relationships with others. Although the many roles it plays are rarely appreciated, it is one of the most remarkable and highly versatile parts of the human body.

Simply put, skin is the flexible, continuous covering of the body that safeguards our internal organs from the external environment. It protects us from attack by physical, chemical, and microbial agents and shields us from most of the harmful rays of the sun, while it works hard to regulate our body temperature. Far from being an impervious barrier, however, the skin is a selectively permeable sheath. It is constantly at work as a watchful sentinel, letting some things in and others out. The skin is also home to hundreds of millions of microorganisms, which feed on its scales and secretions.[1] But our skin is more than a defensive shield, a gatekeeper, and a personal zoo.

The pores and nerve endings of our skin unite us with our surroundings. Skin is the interface through which we touch one another and sense much of our environment. Through our skin, we feel the smooth cold of melting ice, the warm and gentle breeze of a summer evening, the annoying pinch of an insect bite, the humbling pain of a scraped knee, the soft and calming feel of a mother's hand, and the thrill of a lover's touch.

Throughout the approximately six-million-year journey of the human lineage, our skin has traveled and evolved with us, through myriad changes of climate and lifestyle.[2] In addition to providing a boundary layer between the body and the environment, the skin has taken on the new roles of social canvas and embodied metaphor in our recent evolutionary past. Our skin reflects our age, our ancestry, our state of health, our cultural identity, and much of what we want the world to know about us. People in all known cultures modify their skin in some way, often using deliberate marking and manipulation to convey highly personal information about themselves to others.

No other organ in the body can boast so many diverse and important roles. Few people, in fact, think of our skin as an organ of the human body. The word "skin" doesn't bring to mind the same meaty image as the term "liver" or "pancreas," nor does it elicit the same queasy response. Yet the skin is, by most measures, the body's largest organ, and it is certainly the most visible.[3] Its size is about two square meters (approximately twenty square feet), and its average weight is four kilograms (about nine pounds). Unlike a heart or a kidney, skin never fails, because it is constantly being renewed.

Human skin is unique in three respects. First, it is naked and sweaty. Except for the scalp, the groin, the armpits, and the male chin, our bodies are effectively hairless. Because this phenomenon sets us apart so obviously from other mammals, it has engaged the attention of many scientists and armchair theorists, who have produced a welter of unconventional explanations. Among the many theories put forward to account for our hairlessness, the most cogent proposes that we have lost most of our body hair in order to keep cool in hot environments and during exercise. Humans sweat to a greater degree than other mammals, and hairless skin allows sweat to evaporate quickly, more efficiently cooling the body.[4]

The second distinctive attribute of human skin is that it comes naturally in a wide range of colors, from the darkest brown, nearly black, to the palest ivory, nearly white. This exquisite sepia rainbow shades from darkest near the equator to lightest near the poles. This range forms a natural cline, or gradient, that is related primarily to the intensity of the ultraviolet radiation (UVR) that falls on the different latitudes of the earth's surface. Skin color is one of the ways in which evolution has fine-tuned our bodies to the environment, uniting humanity through a palette of adaptation. Unfortunately, skin color has also divided humanity because of its damaging association with concepts of race. The spurious connections made between skin color and social position have riven peoples and countries for centuries.

The third major distinction of human skin is that it is a surface for decoration. Our skin is not just a passive covering that betrays our age or physiological state. It is a potentially ever-changing personal tapestry that tells the world about who we are or who we want to be. And, unlike the involuntary advertising afforded by our own skin color, the decorations we place on our skin are deliberate and willful forms of advertisement—skin becomes a social placard, serving as both our "advertising billboard and the packaging."[5] No other creature exerts such extensive control over what its skin looks like. Humans expose it, cover it, paint it, tattoo it, scar it, and pierce it, telling a unique story about ourselves to those around us. In a world of increasingly globalized fashion, adornment of the skin is one of the last frontiers of individuality and personal adventure.

More than any other part of the body, our skin imbues us with humanity and individuality and forms the centerpiece of the vocabulary of personhood. The word "skin" often represents the whole body or the wholeness of the self, and its use in figures of speech can convey intensely personal feelings or strong sentiments about identity and appearance.[6] Think about the times you use the word "skin" in a figure of speech in normal conversation and how frequently you see it used in writing. It can express surprise or fear—"I nearly jumped out of my skin!" After a close call, you might exclaim, with relief, "I just managed to save my skin"; similarly, the Bible quotes Job as saying, "I am escaped with the skin of my teeth" (Job 19:20). We use the metaphor of skin to describe a person's sensitivity: "It won't bother her—she has a thick skin"; or "It's no wonder she took offense at your remark—she has an incredibly thin skin." We can dismiss others or complain about them with the words "So what if he doesn't want to go? It's no skin off my nose" or "He really gets under my skin." An extremely thin person is "nothing but skin and bones," while an attractive individual is cautioned that beauty is "only skin deep." T. S. Eliot's poem "Whispers of Immortality" opens with this image: "Webster was much possessed by death / And saw the skull beneath the skin."[7] These images and patterns of usage maintain a currency and an immediacy because we so closely associate our skin with the essence of our being. They induce empathy because of the unambiguous association of the skin with a vulnerable self.

As a teacher of human anatomy for many years, I was always fascinated by the reactions of students as they began to dissect a human body at the beginning of the academic year. Most students approached the task with hesitation, and some with great fear. For many, this reluctance was born of the simple dread of touching a dead person, something that many of them had never done before. Much of their reserve, however, derived from a sense of trepidation about trespassing a boundary they had not considered crossing. The intact skin of the cadaver, especially the skin of the face, was associated with a real person who had lived a real life of laughter and tears—a person like them, who had felt joy and sorrow just a few months or years earlier. But as they carefully began to remove the skin, their hesitation and reserve slowly disappeared. Although the body was no deader without skin, the partially flayed cadaver lacked the covering that the students associated with dynamic personhood. The veil of personality and individuality had been removed, revealing the muscles, nerves, and sinews of the human species.

A modern *écorché,* an anatomical representation of the body with the skin removed (see color plate 1), shows us that the body without skin is indeed human, but it has been stripped of its identity and personality. Lacking its inherent color, scars, decoration, and any trace of emotion, the cadaver is a human, but not a person. It beckons us to ponder the meaning of the skin as a barrier and, indeed, to consider our definition of an individual. As the cadaver proffers its skin, it arrests the viewer with the question, "Who am I without my skin?" Without its skin, however, it also portrays a universal humanity that invites us to learn more about our shared history underneath the skin. Only after my students had removed most of the skin from the body could they begin to enjoy the wonders of anatomy without reservation and to explore the body's complex and mysterious interior. They no longer felt that they were violating a person's intimate space.

Despite the importance of skin in human biology and in interactions between people through the ages, most of its rich and interesting history has not been told. This book is an attempt to do so. It is not a systematic treatise or a manual, but more an idiosyncratic guidebook, replete with personal detours into topics about skin that have most engaged me in my work over the years. As an anthropologist with training in comparative biology, I have enjoyed learning about the details of hippo and bat skin as much as I have enjoyed exploring the intricacies of human skin, so be prepared for a few unexpected and unusual facts. The main purpose of this book is to provide information rather than advice, but in a few very important areas—such as sun and skin care, and skin color and race—advice follows naturally from the information about why and how these matters came to be of importance to us today.

The first chapter looks at how skin is put together and how it works. It provides a tour of the basic aspects of the skin's structure and function with the aid of simple illustrations that relate the layered construction of skin to its varied roles. Once these are understood, the many services the skin performs in normal life start to become clear, and you will find yourself appreciating and respecting your skin perhaps more than you did before.

Chapter 2 uses the tools of comparative biology to relate the evolutionary history of skin, a story that spans more than three hundred million years. Examining evidence from comparative anatomy, physiology, and genomics allows us to reconstruct the major steps in the evolution of the skin of land-dwelling vertebrates. It turns out that the fossil record helps us only a little here because traces of skin are rarely preserved for more than a few thousand years, and fossilized skin itself is almost nonexistent. This chapter focuses especially on the evolution of the outermost layer of skin—the epidermis—and how its structure has made life on land possible, for humans and all our terrestrial ancestors.

From this general discussion of vertebrate skin, chapter 3 turns to a more detailed examination of human skin, particularly to the topics of human hairlessness and sweating. Sweating is one of the most important functions carried out by our skin. Although most people in industrialized societies view sweat as undesirable, we would be in very bad shape without it. Sweat helps to cool our bodies, including our heat-sensitive brains, an ability that was indispensable in human evolution.

We can look at the skin as a factory, that is, a place where chemical processes occur, including those generated by sun exposure. Much of what goes on in our skin from a physiological point of view is a response to sunlight and to UVR in particular; different wavelengths, or energy levels, of UVR affect different processes. Chapter 4 tells the important story of vitamin D production in the skin, while chapter 5 specifically deals with the role of melanin, the main pigment in human skin. Melanin has served enormously important functions in biological systems for at least the past four hundred million years, absorbing high-energy solar radiation and protecting our bodies from many of the harmful chemicals such radiation produces.

Chapter 6, on skin color, is closest to my heart and personal interests because I have studied the evolution of skin color in humans for nearly fifteen years. Until fairly recently, little research on the biology of human skin color had been conducted, and neither the scientific community nor the public had much understanding of the topic. In part, human skin color was considered too socially sensitive a topic to be broached in the halls of academe or applied science; skin color remained something that everyone noticed but no one talked about. This situation has changed dramatically in the past decade as new studies have shed light on the evolutionary reasons why people have different skin colors, the genetic variations that contribute to these differences, and the significance that these differences have for people's health and well-being in various environments. Skin color warrants attention because it affects so many aspects of our lives, from our health as individuals to the way we treat each other within societies.

Chapter 7 explores the topics of skin as the organ of touch and as a vehicle for communicating information through tactile sensation. It looks not only at the importance of tactile sensitivity in the evolution of primates but also at the significance of touch in nearly every aspect of our lives, from finding and eating food to communicating our most intimate feelings to one another. This chapter also looks at human fingerprints, from their original significance in our ancestors to the use of modern technology that documents their uniqueness.

Because our skin reflects our emotions and mood states, it can betray our true feelings even when we wish it wouldn't, as chapter 8 points out. The hot, red face of embarrassment and the gray pallor and cold, sweaty palms of anxiety are familiar to all of us and are now well understood. Skin in many animals, including humans, is a natural notice-board that conveys considerable information not only about one's age, health, and emotions but also—in some species—one's degree of sexual receptivity.

Because skin indeed follows "the way of all flesh," chapter 9 explains—without most of the gruesome detail of a dermatology textbook—how our skin reacts to the ravages of age, the environment, and disease. As our "face to the world," the skin sustains injuries at different scales on a daily basis and is subject to infection and infestation by a bewildering array of organisms. This chapter also discusses our modern obsession with age-related changes in the skin, especially wrinkles.

Chapter 10 presents an overview of what humans, as complex cultural beings, do to our skin, a topic that has been the focus of many recent books.[8] To appreciate the origins and great antiquity of skin decoration, we begin with a short history of the subject. As we survey various types of skin marking and trends in skin modification, you will discover that certain themes recur across cultures and through time. In particular, this chapter draws attention to how people have used cosmetics and paints to establish their identity and advertise their sexuality and explains how tattoos have served as ways of expressing either individuality or group affiliation in human societies.

The book's final chapter looks at future skin, from the prospects for custom-made artificial skin that can be used in clinical contexts to the expanding frontiers of communication and entertainment via remote touch. Skin and tactile communication have always been important to people, but they will soon become ever more so as we develop more sophisticated means for conveying and detecting remote touch. We are entering another brave new world in which skin will play new roles in interpersonal and broader social communication. For much of modern human history, skin has been a canvas for human creativity, setting us apart from our primate relatives with a mantle of uniqueness. So it is likely to remain.

1
skin laid bare

It isn't good to take for granted something as important as skin. Take a moment and imagine the following scene. You're standing in the moist, shadowy heat of an orchard in the late afternoon of a summer's day. You are able to stand outside in comfort without overheating, thanks to your skin's ability to regulate your body temperature and shield you from ultraviolet radiation. Only a few beads of sweat on your brow and upper lip betray the fact that your skin is working to keep you cool. As you flick away the fly that tried to settle on your face, you don't give a thought to the way your skin is protecting you from the microorganisms on the insect's feet and snout.

You have your eye on a peach dangling from a branch above your head, and you want to pick it and eat it. As you reach up toward that lovely peach, you're distracted again by the fly, and the back of your hand scrapes against the snag of an old branch. Thanks to your skin's fairly tough surface, the scrape isn't a problem. A welt starts to rise in a few minutes, but your skin is unbroken because its outermost layer is quite scuff-resistant. You reach up again, and the elastic properties of the skin on your arm and trunk allow you to stretch effortlessly until, on tiptoe, you touch the peach. As you grasp the fruit, you squeeze it ever so slightly and register its subtle softness through the exquisitely sensitive pressure sensors in the skin of your fingertips. It is ripe. As you pull the peach off the tree, the temperature sensors in the skin of your hand let you appreciate its slight warmth. As you lower your arm, the stretched skin of your arm and trunk returns instantly to its resting shape.

You bring the peach to your nose and smell it, and then brush it gently against your cheek, enjoying the feeling of the soft fuzz against your face. Your sensitive facial skin, with its high density of delicate touch sensors, is transmitting information about the texture of the peach to your brain. Just as you prepare to bite into the fruit, an annoying tickle at your ankle disturbs your reverie, and you realize that a mosquito has just bitten you while you were so pleasantly distracted with the smell and feel of the peach.

Your skin and its wide-ranging capabilities made the various parts of this scenario possible. To understand how this is so, an introductory tour of human skin, exploring its structure and its essential functions, is in order.

One of the most striking features of human skin is that it is basically naked; in this way it differs from the skin of most of our warm-blooded relatives. The ancestors of birds and mammals evolved fine, threadlike appendages on their skin—feathers and hairs, respectively, which regulate heat interchange and also help to prevent water loss and mechanical trauma. Lacking such protection, human skin had to undergo numerous structural changes to give it strength, resilience, and sensitivity.[1] Our skin is not perfect, but it does a remarkably good job. Our fabric doesn't wear out, our seams don't burst, we don't spontaneously sprout leaks, and we don't expand like water balloons when we sit in the bathtub.

Some of the most important properties of skin are related to sunlight. In humans, the skin and the pigments it contains selectively filter the ultraviolet radiation emanating from the sun. Our skin has the amazing ability not only to serve as a protective shield against the damaging effects of sunlight but also to utilize some of that same sunlight to the body's advantage, by beginning the process of producing vitamin D right there in the skin. Thus our skin, like so many other parts of the body, is a compromise hammered out at the negotiating table of evolution. Its complex properties reflect a balance, brought about through natural selection, between conflicting needs—in this case, protection against harmful solar radiation and production of an essential vitamin.

Skin is made up of layers with different physical and chemical properties. This laminar, or layered, construction gives the skin its resistance to abrasions and punctures and allows it to avoid absorbing most substances. The skin's two major layers, the epidermis and the dermis, differ remarkably in their composition and function (figure 1). The skin also includes special types of cells that insinuate themselves into the skin during early embryonic development. These aptly named immigrant cells play varied and important roles in protecting the skin, as we'll see later in the chapter.

The skin's outermost layer, the epidermis, shields us from environmental oxidants and heat, while it also resists water, abrasion, stains, microbes, and many chemicals—a list of qualities that makes the epidermis sound more like a revolutionary new type of carpeting than a natural material. It is all the more astonishing, then, that these useful attributes are found in a self-renewing layer only about one millimeter thick, which continuously performs all its functions despite being in a constant state of turnover, with its outermost cells being shed as they are replaced from below.[2] The epidermis is composed mostly of a specialized type of epithelium consisting of multiple layers, or strata, of flattened cells. (An epithelium is a covering of any external or internal surface of the body.) Because these cells contain high concentrations of the protective protein keratin, this epithelium is known scientifically as stratified keratinizing epithelium.

The very surface of the epidermis is its most remarkable layer, the stratum corneum (figure 2). The stratum corneum is sometimes called the epidermal horny layer because it consists of a relatively thin sheet of dead, flattened cells with a smooth, fairly tough, and water-resistant surface. The only things that interrupt its surface are hair follicles, the pores of sweat

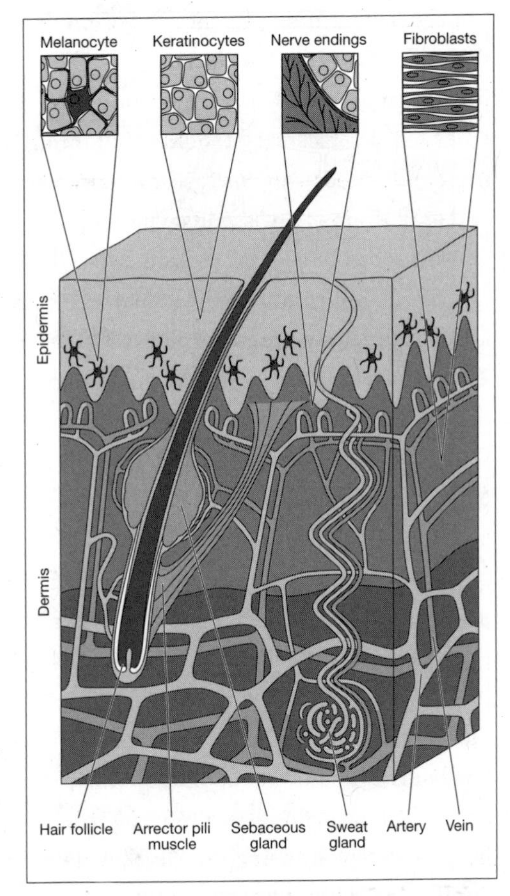

FIGURE 1. A typical section through human skin, showing its laminar construction and some of the glands and immigrant cells that are most important to its function. (© 2005 Jennifer Kane.)

glands, and parts of some of the so-called immigrant cells that help to form the complex mosaic of the skin. The skin's effectiveness as a barrier against environmental insult of all kinds, especially oxidative stress such as ultraviolet radiation (UVR), ozone, air pollution, pathological microorganisms, chemical oxidants, and topically applied drugs, depends primarily on the integrity of the stratum corneum.[3]

One of the ways the skin defends itself against some environmental stressors is to become thicker. When the skin is repeatedly exposed to UVR, for instance, cell division increases in the deepest layer of the epidermis, the

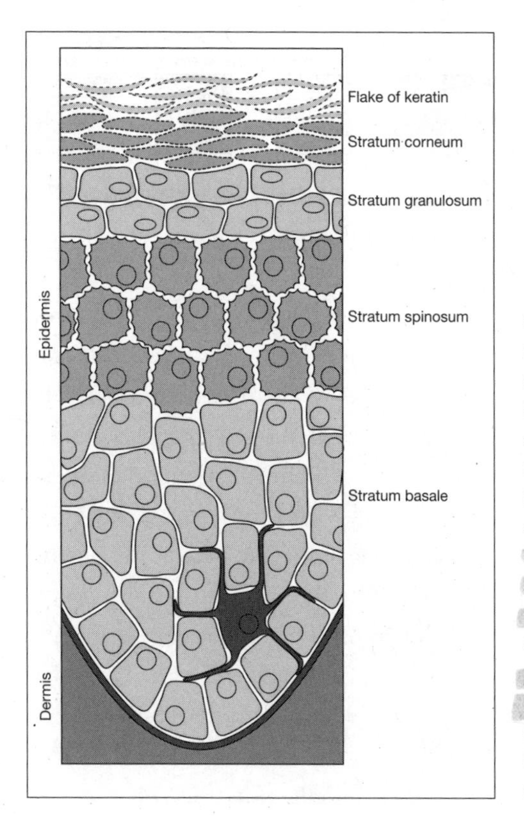

FIGURE 2. Human epidermis, with its four constituent layers and a single, spidery melanocyte near its interface with the dermis. The continuous production of new keratinocytes in the stratum basale helps to maintain the epidermis in good condition. Filaments of keratin that provide structural support and abrasion resistance are produced in the cells while they are in the stratum spinosum. In the stratum granulosum, the cells—with visible internal granules—die and move into the stratum corneum. (© 2005 Jennifer Kane.)

stratum basale, which is the source of epidermal cells; and, as a result, the stratum corneum thickens.[4] If the stress, whether external or internal, is extreme—too much UVR, too much heat, a corrosive chemical such as acid, some diseases or genetic problems—the stratum corneum can cease to be an effective barrier. This can have disastrous results if a large area of the skin is affected.

Keratinocytes, the main types of cells found in the epidermis, are made up of proteins called keratins. They are responsible for the strength, resistance, and stretchability of the skin's surface. Within keratinocytes, filaments of keratin are embedded in a gelatin-like matrix, and layer after layer of these cells build up from below to make up the epidermis. Between the cells, a substance rich in proteins and lipids fills the narrow spaces. The elasticity and imperviousness of the epidermis, especially the stratum corneum, result from its "brick and mortar" construction, that is, the tight and strong physical interconnections between adjacent cells and the protein and lipid material between them.[5] In people with dark skin, the keratinocytes also contain flecks of the pigment melanin ("melanin dust"), which provide another layer of protection against UVR.

Scientists have long thought that human epidermis is unique because it does such a good job of protecting us even though we are effectively hairless. But the genetic basis for that uniqueness had not been appreciated until the past few years. One of the ways in which the genetic makeup of humans varies from that of our closest relatives, chimpanzees, is in the genes determining the structure of the epidermis. The recent sequencing of the chimpanzee genome has revealed that one of the few areas of the genome where humans and chimps differ significantly is in a cluster of functionally related genes that regulate the differentiation of the epidermis and contribute to coding the proteins that make up the keratin-rich layer of the skin.[6] At least as far as primate skin goes, our epidermis is tough stuff.

The immigrant cells in the epidermis are a diverse lot that work with the other cells in the skin. They migrate into the skin from other parts of the body during early development to provide special physical and chemical protection against potent environmental agents such as UVR, disease-causing microorganisms, and dangerously high physical pressures. Although they are developmental interlopers, the immigrant cells don't in any way weaken the physical fabric of the skin. There are three main types of immigrant cells in the epidermis. Melanocytes (shown in figures 1 and 2) produce the skin's primary pigment and natural sunscreen, melanin. These cells migrate to the skin from a position flanking the spine during early embryonic development. Once they arrive, they set up shop near the interface of the dermis and the epidermis in order to manufacture melanin. Some people produce a lot of melanin in their melanocytes, whereas others produce only a little, depending on the amount of UVR present in the environment of their ancestors. Skin color, which is determined by the activity of melanocytes and their manufacture of melanin, has evolved under the close watch of natural selection.

Two other types of immigrant cells are also important. Langerhans cells are specialized cells of the immune system that respond to foreign substances coming in contact with the skin. They constitute the body's first line of defense against bacteria and viruses that land on the skin. Merkel cells are associated with the ends of sensory nerves in the skin, where they appear to assist in the transfer of mechanical signals from the skin to sensory nerves and then on to the brain. Merkel cells, which are common on the smooth skin of our fingertips and lips, contribute to our finely discriminating sense of touch. They are also of great importance to our furred and feathered relatives: in mammals and birds, Merkel cells occur in the collars of cells that support hair and feather follicles, including those surrounding the sensitive whiskers of dogs, cats, and rats.

Probing beneath the epidermis, we reach the second of the skin's two primary layers, a thick layer of dense connective tissue called the dermis. This is the layer that really imparts toughness to skin. It is pliable, elastic, and has considerable tensile strength. Most of the thickness of our own skin—and most of the thickness of the hide of any animal—comes from the dermis.[7] Its thickness, in addition to its chemical and physical properties, helps to insulate the body and makes the skin resistant to mechanical injury. Leather is composed mainly of tough animal dermis that has been tanned so that it will be more pliable.[8]

The dermis is a composite tissue that gets its strength and toughness from a combination of collagen fibers and elastin fibers. These fibers are maintained in a gel composed of salts, water, and large protein molecules called glycosaminoglycans. The primary cells of the dermis are collagen-rich cells known as fibroblasts. Collagen, which constitutes 77 percent of the dry weight of skin, accounts for most of the tensile strength of the skin and for some of its ability to scatter visible light (figure 3). Collagen acts just the way it looks, like tough little ropes of protein holding the dermis

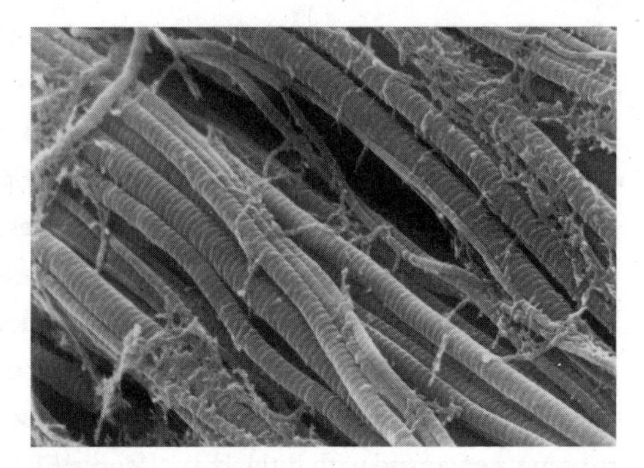

FIGURE 3. The collagen network in human dermis. The dermis contains several types of collagen arrayed in a dense mesh that maintains the physical integrity of the skin. (© 2005 L'Oréal Recherche.)

together. Interwoven with the collagen is a network of abundant elastin fibers that restore the skin to its normal configuration after stretching.

The production of collagen and elastin fibers slows down as we get older, and it is adversely affected by UVR from excessive sun exposure. Many products on the beauty market today claim to stimulate production of these materials to keep skin looking young. But there is only so much that creams, treatments, and "cosmeceuticals" can do to change the appearance and composition of skin, especially when people have caused irreparable damage through their incautious behavior in the sun. Many of the processes in the skin that control the production of collagen and elastin are governed by internal mechanisms of cellular aging that are not affected or are only weakly affected by what we apply to the skin's surface.

Amid the complex tangle of connective tissue fibers in the dermis, we find a branching network of blood vessels, an extensive network of nerves, numerous sweat glands, and an assortment of hair follicles, hair-raising arrector pili muscles, and oil-producing glands (refer back to figure 1). The blood vessels are critical because they supply the appetites of the sweat glands, the hair follicles, and the rapidly multiplying cells in the lowest layer of the epidermis. The density of blood vessels varies over the body's surface. They are especially concentrated on the head, for instance, where temperature regulation is particularly important to protect the brain and where the hair follicles of the scalp require good nutrition from a rich blood supply so that hair can grow. Blood vessels are also quite dense in areas where the skin must be kept moist by sweat and sebaceous (oil-producing) glands—for example, on the palms of the hands, the soles of the feet, and the nipples. In addition, blood vessel density is related to different postures. In both humans and primates, some of the densest concentrations of blood vessels in the body are found on the bottom of the buttocks, supplying the skin in this area with blood so that it does not deteriorate when we sit for long periods. In some of our close primate relatives, the skin around the female genitals is richly supplied with blood vessels, which permit the skin to become engorged with fluid when the animals are sexually receptive, creating puffy pink sexual swellings that are highly attractive to males.

The blood vessels of the dermis carry red blood cells, which derive their color from hemoglobin. Hemoglobin is a pigment that is bright red when it is carrying oxygen to cells and a dull reddish-blue after it has discharged its ferried oxygen and is heading back to the heart and lungs. Hemoglobin is one of the skin's main pigments, but it is most visible in people who have relatively little of the dark brown melanin pigment in their skin. Rosy cheeks and blue veins are more evident in people with light skin than in those with dark skin. The painfully bright red appearance of sunburned skin actually results from an increase in the number and diameter of the tiny blood vessels in the skin as well as an increase in the blood flow through each of these vessels. Sunburned skin feels hot to the touch because it is infused with blood and because it is mounting a hot and vigorous inflammatory response in order to repair the damage caused by UVR.

The nerves of the dermis are highly complex because the skin is one of the body's main sensory portals. Skin contains several specialized types of receptor cells, which send signals to the central nervous system about the external environment and the state of the skin. These include two types of temperature receptors, diverse mechanical receptors associated with both hairy and smooth skin, and an important group of pain sensors that specialize in detecting potentially dangerous physical stimuli or the presence of injury or inflammation. Although this formidable battery of receptor cells is extremely important, their evolutionary history is not yet well known.

A tour of the skin would not be complete without a side trip to examine hair. Hair in humans is significant largely because we have so little of it. If we cast our gaze back in time to consider the evolution of skin in our earliest warm-blooded ancestors and cousins, the story of hair becomes very interesting. As the forebears of mammals and birds evolved toward endothermy, or warm-bloodedness, one of the key innovations that allowed this development was good external insulation on the body. In other words, if you want a warm house, but not a high heating bill, you must have good insulation in your walls. Warm bodies permitted higher activity levels throughout the day, but at the cost of greatly increased energy expenditure. In the ancient physiological economy of proto-birds and proto-mammals, keeping the lid on energy costs was a high priority so that the animals would not have to spend excessive amounts of time finding and eating food. The solution was found in the development of complex, built-in insulation such as hair and feathers.

The ancestors of mammals and birds possessed follicles in the skin, from which hair or feathers could develop. Follicles contain protected collections of special generative cells that allow feathers or hair to grow from within the skin.[9] These cells are a type of stem cell unique to the epidermis, and they maintain the hair follicle and regulate the cycle of hair growth. The hair follicles of mammals differ from the follicles of birds, and even among mammals different types of follicles are found in the body. Oddly, one of the most highly specialized is the mammary gland. In mammals, mammary glands develop as fantastic elaborations of the branching processes within very specific follicles of the chest wall. Milk is produced in the coils of the gland when the appropriate hormonal signals are received. Mammary glands are a delightful example of how evolution works on existing structures in order to make something new. In this case, retooled follicles developed into an effective means for nurturing mammalian infants.

In humans, numerous hairs grow from hair follicles located in the der-

FIGURE 4. Piloerection, literally a "hair-raising" reaction. Piloerection is one of the ways in which an agitated or angry mammal can make its feelings known. Here, a chimpanzee makes a threatening display to members of its group. (Photograph courtesy of Frans de Waal.)

mis. We actually have just as many hairs on our bodies as monkeys and apes do, but ours are much thinner and, on some parts of the body, nearly invisible. Hair follicles are important placeholders in human skin and are associated with complex sensory receptors and sebaceous glands. Even though we don't really need the hair anymore, hair follicles serve as organizing centers within the skin. The twitching whiskers of cats and mice are highly specialized hairs called vibrissae whose follicles are richly supplied with nerve receptors (the Merkel cells mentioned earlier), which convey detailed information to the animal's brain about what it is touching. We and our closest primate relatives don't have vibrissae on our snouts, but we have quintessentially sensitive hands that do the same job.

Hair performs a wide range of functions in mammals. Ironically, given how much attention and money humans lavish on hair, it is probably much less important to our survival than it is for other species. For most species, hair insulates, protects against the sun, enhances the sense of touch, serves as ornamentation, and acts to communicate emotion. We don't usually think about hair expressing our emotions, but in many species it does, through the mechanism of piloerection—literally, raising of the hairs. Piloerection occurs when an animal is angry, frightened, or thrilled and serves to make the animal look larger and more threatening (figure 4).

One of the interesting consequences of human hairlessness is that we lack the ability to convey anger, excitement, or fear through highly visible

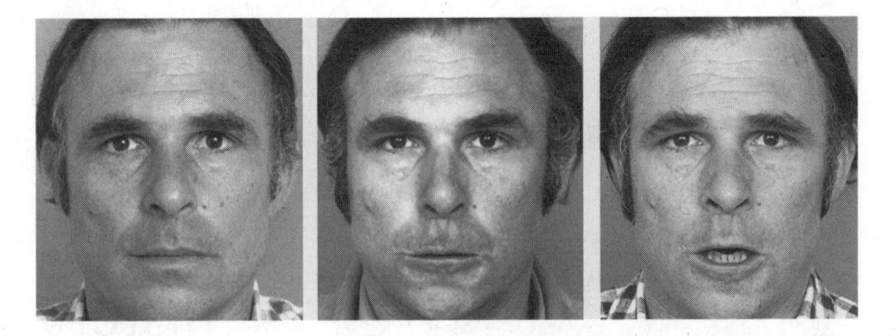

FIGURE 5. Lacking visible body hair, humans can't use piloerection to communicate their anger or fear to others. We compensate for this in part by highly visible facial expressions that are universally understood. The human face on the left exhibits a neutral expression, but those in the center and on the right show incipient or contained anger. Narrowing the lips is a reliable sign of anger in humans. (© 2003 Paul Ekman. *Emotions Revealed.* Time Books.)

piloerection. We often describe fearful situations by saying that the experience was "hair-raising" or that it "made my hair stand on end." When someone angers or irritates us, it's common to say that the person "makes the hair stand up on the back of my neck." At times like these, we may really feel and see the hairs rising on our own bodies. This is exactly what's happening, thanks to our little arrector pili muscles, but most people looking at us can't see that reaction. How do we then visibly communicate these critical emotions? This is another untold and important story of human evolution. When our covering of body hair became less obvious, we had to evolve different, highly visible means to convey our emotions. Part of the solution was our repertoire of facial expressions, which are the most complex and varied in the animal realm (figure 5).[10] Our sensitively expressive faces permit us to convey subtle nuances of information about what we are feeling. Through these expressions, we not only have compensated for lacking body hair that can fluff and bristle but also have developed ways to convey even more information.

2
history

Given the importance of our skin and the functions it performs, it is striking that relatively few scientists study its evolution. Until about thirty years ago, most scientific studies of skin were confined to describing the anatomical details of modern human skin and discussing various skin diseases and conditions. In recent decades, as life expectancies have increased in industrialized nations, with a concomitant emphasis on looking youthful, more research has been concerned with improving the appearance of skin and the effects of topical preparations, injections, and surgery. Only in the past decade have researchers focused on some of the most interesting and fundamental questions about skin, including the evolution of its special protective properties and its various appendages such as hair and nails. With new tools in comparative biology, especially in comparative and functional genomics, now available, questions about the evolution of skin that were once considered too difficult or even unanswerable can be tackled.

Some of the most important phenomena in evolution—short-lived behaviors, for instance, or perishable tissues like skin—die with the animals to which they belonged, leaving no trace. Studying these developments is worthwhile, however, even if it is difficult, because temporary structures or behaviors can sometimes hold the key to understanding why an organism was able to survive or reproduce successfully. New insights from evolutionary theory, combined with the introduction of new investigatory methods, are now making it possible to capture such lost and valuable pieces of information, which may lie outside the fossil record.

How can we investigate the evolutionary history of a part of the body that is almost never preserved in the fossil record? Like other soft tissues of the body, skin generally does not last long after death. It hardly ever leaves an imprint or becomes petrified, as a bone or a tooth might. In the study of fossils (paleontology), researchers can often derive information about some of the body's soft tissues, like muscles and ligaments, indirectly. Because these tissues attach directly to bones, they sometimes leave a trace of their attachment on a bone's surface, providing clues about their size and structure. From such evidence, we might be able to discern the overall shape of an animal or how it moved or how it ate its food.[1] Bones also contain holes through which nerves and blood vessels passed during life. These can allow us to make inferences about the size and importance of the nerves and vessels based on the size of the holes they left behind. But reconstructing anything about skin from fossil bones is almost impossible—because skin doesn't attach directly to bone, it leaves no such clues.

When it comes to records of ancient skin, the best we can usually hope for are imprints of skin that are preserved as fossils. Animal and human footprints are made by skin-covered feet, but because the imprints are made in sand, volcanic ash, or mud, they are often too blurry to tell us much about the skin itself.[2] Rather, fossil footprints and trackways have proven more useful for reconstructing details of how animals moved. There are, however, two well-known cases of fossilized skin, both involving dinosaurs. In the first, a piece of the skin of "Leonardo," a young adult ornithopod di-

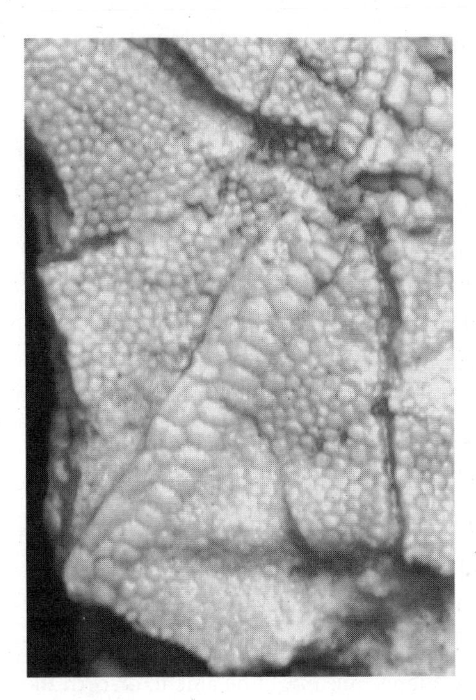

FIGURE 6. This remarkable photograph captures the surface detail of the skin of an embryonic sauropod dinosaur. What appears to be preserved skin is actually a natural cast of the original integument. A close look reveals the small, overlapping, tubercle-like scales that covered the surface of the embryo's skin. (© Lorraine Meeker. Photograph courtesy of Luis Chiappe, Natural History Museum of Los Angeles County.)

nosaur belonging to the species *Brachylophosaurus canadensis,* was preserved when the animal was naturally mummified and then buried by sediment. In another rare and fascinating example, natural casts of the skin of embryonic sauropod dinosaurs were discovered inside the remains of their eggs (figure 6).[3] The exceptional circumstances under which these specimens were preserved allowed the fine details of the skin to be faithfully maintained, leaving us with a singularly beautiful and accurate record of the texture and shape of the animal's skin.

Rarely, we may find the skin of an animal or person who died within the past few thousand years and was preserved under certain exceptional physical or chemical conditions. The familiar mummies of ancient Egypt, for example, were carefully dried with natron, a carbonate-based compound, and then treated with a suite of other chemicals in order to maintain them—skin and all—for a long and noble afterlife. Less well known are the naturally occurring mummies of humans and animals found in dry and protected conditions: in a high, cold desert; in the caves or secluded recesses of high mountains; or in the hot, drying conditions of a low-lying desert with good air flow. Under such conditions, a body will dry out (or, in some cases, freeze-dry) faster than it can decompose, leaving at least some of its skin intact.[4] Hot, dry, and windy conditions led to the preservation of the famed mummies of the Silk Road in the Tarim, Turfan, and Hami basins of far western China over the past four thousand years. The skin of these mummies is sufficiently well preserved that one can recognize their relatively fair complexions and European-looking features. This evidence, combined with their distinctive clothes, alerted researchers to the fact that these ancient inhabitants of what is today Xinjiang Province had migrated there from the Caucasus region of western Asia during a period when the basins of Xinjiang were more hospitable than they are today.[5]

On even rarer occasions, bodies may be fast-frozen after being covered with snow or ice immediately after death—for example, if they were entombed by an avalanche. Under these conditions, bodies, including skin, are preserved fresh, as they would be in a freezer. This was the fate of a Late Pleistocene baby mammoth, nicknamed "Dima," who was found in frozen peat tundra near the Kolyma River gold placer in Siberia in 1977 (figure 7). "Ötzi," a Neolithic-age "Iceman" whose frozen body was found by alpine hikers on the border between Italy and Austria in 1991, was also the victim of an avalanche (see color plate 2). In both cases, the skin was fairly intact at the time the bodies were discovered and began to deteriorate only when the bodies were exposed to air and warmer temperatures. Since the recovery of Ötzi's body, it has been maintained under constant refrigeration and examined in detail by teams of doctors, anthropologists, and forensic scientists.[6]

Even rarer than mummification and freezing is the preservation of skin in the highly acidic conditions of a peat bog. The water in peat bogs is cold and acidic, inhibiting the growth of microorganisms that could cause decay. Such were the conditions under which the bodies of the so-called bog

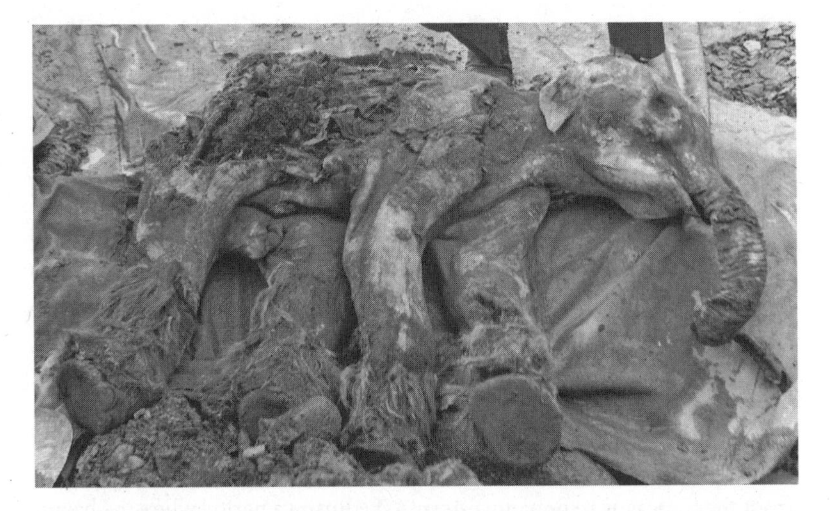

FIGURE 7. Frozen in the Siberian tundra, the skin of the baby mammoth Dima was extremely well preserved. (Photograph courtesy of Anatoly Lozhkin.)

people were discovered in several of the most famous peat deposits in England, Denmark, and Germany (see color plate 3). Like mummies, the bodies of bog people are a few thousand years old.

These discoveries are remarkable and interesting, but none of these human or animal remains are really very old in evolutionary terms. The oldest known preserved skin, that of Ötzi, the ancient Tyrolean, is only about five thousand years old and is completely modern in appearance and function. Neither Ötzi, nor Dima, nor any of these other colorful or curious characters from the recent past tell us anything about the truly ancient history of skin. When and how did skin evolve in the first place? What did the skin of the first land-living animals look like? How and when did human skin come to look, feel, and function as it does today?

Scientists believe that the earliest multicellular organisms, which lived in the sea, consisted of colonies of single types of cells. The surface of the cells that faced the open water often sported a thickened membrane as well as small tail- or whip-like structures that allowed the organism to move around to a limited extent. When such organisms began to get larger and became more solidly filled with cells, several very important things happened. Because the cells on the outside could not absorb enough nutrition and oxygen through their membranes to support the needs of the entire organism, a new structure evolved to permit food and oxygen to get inside. This opening was a simple mouth, which led to a pocket lined with specialized cells. Through this lining—a primitive type of epithelium—dissolved nutrients and oxygen from the marine environment could be absorbed and distributed to waiting cells inside. This was the first pharynx, or primitive gullet.

As these early organisms got even bigger, the problem became one of distribution. How could dissolved nutrients and oxygen be delivered efficiently to the cells that needed them? To meet the metabolic needs of larger organisms, different types of cells—the earliest tissues—began to evolve. It is at this point that we witness the evolution of tissues with specialized functions, such as sensory cells and nerve cells.

With increased complexity on the inside, the organism also needed better protection from the environment outside. In the most primitive invertebrates (animals without backbones) living today, such as sponges and jellyfish, we find undifferentiated protective layers of cells on the outside of the organisms. More advanced invertebrates, such as the widely studied nematode *Caenorhabditis elegans,* have a simple epidermis that acts as a shield against potential disease-causing microorganisms—it keeps the outside out.[7] In this way, a different chemical environment—tuned to the needs of specialized internal body tissues—can be continuously maintained inside the body. The epidermis of invertebrates is not skin as we usually think of it, because it is composed of only one layer. Yet, in its structure and function, it is the primitive forerunner of our skin and that of all other vertebrate animals (those with backbones).

Most of what we know about the skin of the earliest vertebrates we have inferred from studying the epidermis of some of the most primitive living vertebrate creatures, such as certain kinds of fish and aquatic amphibians. The epidermis of these animals must be protective, sensitive, and capable of negotiating complex biochemical interactions between the organism and its environment. The most critical of these interactions is the movement of various substances, including salts, water, and oxygen, across the skin so that the animal can maintain a constant inner environment.

Because it must perform so many functions, it's not surprising that the skin of even the most primitive vertebrates contains a multiplicity of cell types. Anyone who has ever picked up a fish knows that its surface feels smooth, sometimes even slimy. This is because the surface cells in the epidermis secrete mucous, in one-celled mucous glands.[8] The secretion creates a layer called a mucous cuticle, which reduces drag on the fish's skin as it swims and is thick and fibrous in spots, in order to guard the delicate inner organs of the fish against potential abrasion. Fish skin is also penetrated by nerve fibers, many of which are associated with specialized collections of sensory cells, including external taste buds. And when we look into the gills of the fish, we find an even more specialized covering, an epithelium designed to transport oxygen, chloride, and other small molecules and ions. By regulating the passage of these materials, the gill epithelium allows the fish to maintain a proper electrolyte balance and oxygen supply inside its body.

When the first vertebrates crawled onto dry land for the first time, they (and their skin) faced formidable challenges. For these first tetrapods, as they are called, the transition to life on land required many anatomical and physiological modifications, including the loss of external gills. Their bodies, of course, still needed dissolved oxygen and essential salts. At this point, their outer body covering began to change so that skin on the surface of the whole body could take over some of the functions of ion and water regulation performed by the specialized gill epithelium in fish. When we look at the descendants of these earliest tetrapods today—living amphibians—we find telling features in their skin that reflect a shift toward a life spent at least partly out of water. Like fish, amphibians produce mucous in their skin to prevent desiccation, but their mucous-producing glands are multicellular rather than one-celled. Specialized cells in the skin, known as flask cells, help the animals to maintain proper concentrations of salt and water internally.

All amphibians have keratin in the external layer of their epidermis. Keratin is not present in the skin of most fish, but it is found in the skin of all adult amphibians and in all reptiles, birds, and mammals. Keratin protein complexes are the main ingredients of the dead, impervious cells that form the outermost layer of the epidermis. Keratins, in their two major forms of α- and β-keratin, have been around for a long time in vertebrate evolution. Keratinized structures play two main roles.[9] First, they function as tough appendages of the skin—for example, the breeding tubercles in fish, feathers in birds, hairs in mammals, and the many shapes of claws and nails sported by most tetrapods. Second, they serve as a mechanical reinforcement in the skin to provide additional protection against abrasion and as a passive barrier to water movement. Both functions are critical for animals who spend more time out of the water than in it. For amphibians, both mucous secretion and keratinization of the surface layers of skin protect against attacks by microorganisms. Additionally, in some amphibians, granular glands produce various venoms and irritants to further guard against infection and render the animals poisonous or distasteful to predators. When you are a small animal lacking big teeth or the means of making a swift getaway, it pays to taste bad. Toads have become infamous for producing irritating substances on their skin, but many other amphibians produce them as well.[10]

Among the ancestors of the earliest reptiles, the adaptation to full-time life on land was accompanied by drastic changes in the structure of the skin. In these animals, lungs had completely taken over the skin's role of harvesting oxygen from the environment, and kidneys were doing the job of maintaining the body's proper internal salt balance. But the animal still needed to avoid drying out in the open air and getting badly scuffed up as it moved around. The solution to these problems was probably the single most significant step in the evolution of tetrapod skin: the origin of the stratum corneum. For early reptiles, this development was nothing less than

FIGURE 8. In snakes, such as this eastern indigo snake *(Drymarchon couperi)*, ecdysis, or shedding of the skin, begins at the head and continues down the length of the body over the course of several minutes or hours. Sometimes the snake will rub itself against vegetation or rocks to help loosen its old skin. (Photograph courtesy of Kira Od.)

revolutionary. It prevented water loss from the skin and provided enhanced mechanical protection through layers of flattened epithelial cells composed of matrix proteins, all-important keratin, and complex lipids.

To counter the stiffening effect of keratin on the stratum corneum and to retain their bodies' flexibility, reptiles evolved scales. Scales are an engineer's dream: beautiful, economical of materials, and superbly functional. Scales are basically pleats in the skin formed by a protruding outer layer and a soft inner layer that acts as a hinge.[11] This great design has a problem, however. Once formed, individual scales can't grow cell by cell as the animal grows larger. Thus, to accommodate growth, it must form a new set of scales to replace the old set. Lizards and snakes, for example, renew their scaly costume all at once in a cyclical process called shedding, or ecdysis (figure 8). In a synchronized sequence, the animal forms an entirely new epidermis, the layer of older epidermis separates from the new, and the older layer falls away in one piece or several large pieces.

A life spent entirely on dry land requires that the skin not only prevent the underlying tissues from drying out but also protect the body from abrasion. Keratin helps, but those reptiles who live very long lives and endure much wear and tear on land need something even tougher. In the skin of crocodiles and alligators, ossifications—actual bones—evolved in the dermis to ensure abrasion resistance and enhance the animals' defenses against predators. These dermal bones are what give these animals

FIGURE 9. The shell of the giant Galápagos tortoise is a composite structure derived from the animal's spine, ribs, dermis, and epidermis. The remarkable toughness of the shell owes much to the β-keratin present in the epidermal layer. (Courtesy of Bonnie Warren, California Academy of Sciences.)

an irregular, corrugated skin.[12] Independently, shells evolved in the turtle lineage. The turtle shell incorporates the animal's spine, ribs, dermis, and an outer keratinized epidermal layer in a remarkably protective box that grows by slow accretion as the animal's size increases throughout its life (figure 9).

In the reptilian lineage leading to birds, many remarkable things happened, and some of the most important had to do with skin. Compared with the skin of reptiles, the skin of birds shows much greater diversity. Bird skin sports several kinds of scales and, of course, feathers. Feathers are the most interesting and complex of all the so-called appendages of the skin, facilitating insulation, communication, and flight. The evolution of feathers provides one of the most interesting episodes in all of vertebrate evolution.[13] As the ancestors of birds and mammals evolved toward warm-bloodedness, heat-preserving appendages—feathers and hair—evolved from the scales of their respective reptilian forebears. Feathers grow the same way hair does, from follicles containing well-protected stem cells in the skin. While hairs grow as single filaments from follicles, feathers develop from complexly branched filaments. The earliest feathers had a relatively simple structure and apparently helped to insulate the first birds against heat loss. As feather filaments evolved, the branching process became more elaborate, leading to the feathers of modern birds and—eventually—to specialization that allowed flight.[14]

Thus far, creatures adapting to life on land had evolved structures in the skin to protect against desiccation, abrasion, and—in warm-blooded birds and mammals—heat loss. But, eventually, that wasn't enough. Animals on the move on dry land needed help to travel over the varied surfaces of solid ground and vegetation, which weren't always horizontal. For many tetrapods, the answer was, literally, "Get a grip." Claws are specialized modifications of the skin at the ends of appendages that help to ensure secure footing on surfaces of varying angles and textures. Some amphibians, most reptiles, and all birds have them. Claws started out as special keratinized thickenings of the epidermis at the ends of fingers and toes. They are made from wear-resistant β-keratin, and their shape reflects their function: animals with flat claws tend to live on the ground, whereas those with curved claws tend to live in trees. Like other tetrapods, most mammals have claws, but some have evolved modified versions, in the form of nails and hooves, to better suit their requirements for tactile sensation and movement.

Nails are one of the features that distinguish primates from other mammals. The evolution and structure of nails in primates appear to be related to heightened sensitivity on the pads at the ends of the fingers and toes. A discriminating and precise sense of touch (a topic we'll revisit in chapter 7) is essential for life as a primate. All primates, including humans, use their fingers and toes to choose and pick appropriate food and to communicate with one another through grooming and other reassuring gestures. In an interesting twist of evolution, prosimian, or so-called lower, primates such as the slow loris have reevolved a single clawlike nail, known as a grooming or toilet claw, on their index fingers to aid in grooming (figure 10). Humans use our nails for grooming, too, in a variety of ways familiar to us all, some socially acceptable and others repugnant. But we also have made

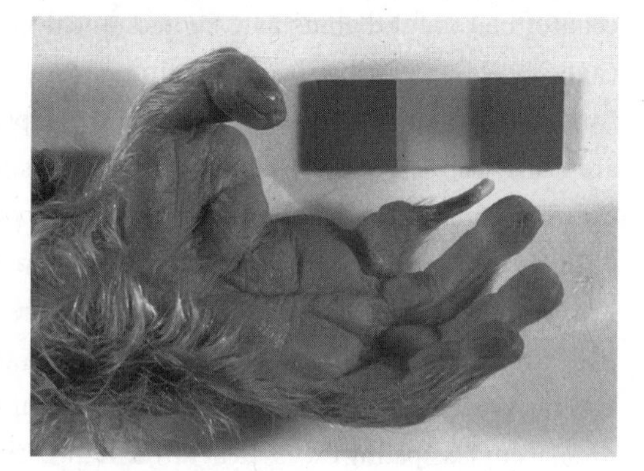

FIGURE 10. Primates' nails provide physical support and protection to the large, sensitive, and vulnerable ends of the fingers and toes. The hand of the slow loris, *Nycticebus tardigradus*, sports a modified nail, or toilet claw, used for grooming. (Photograph courtesy of Vernon Weitzel.)

nails into miniature galleries of self-expression by decorating them in elaborate and colorful ways.

In the lineages of mammals that evolved limbs adapted to long-distance travel on land, claws changed in ways that prevented excessive wear and tear at the ends of the fore- and hindlimbs. The hooves of deer, zebras, and domestic animals such as cows and horses are simply exaggerated claws composed of tough β-keratin that wrap around the ends of the limbs, providing a durable interface between the tips of their toes and the ground. The general name given to these animals—ungulates—is in fact derived from the Latin word for hoof, *unguis*. In every mammalian lineage, the skin and its various appendages have taken slightly different evolutionary paths, as dictated by natural selection. Once a lineage starts in a particular direction—as with the evolution of hooves—there is no going back. In the case of hooves, wear resistance came at the price of sensitivity. A horse may be able to run fast and long on the reinforced tips of its toes, but it can't squeeze a ripe fig or play the piano.

We have seen that the skin in different lineages of tetrapods evolved in several different directions, especially when it comes to hair, feathers, and nails. But what happens when natural selection favors the loss of these structures? This problem has arisen repeatedly in mammalian evolution, particularly in relation to hair. The vast majority of mammals bear handsome coats of hair or fur that insulate, protect, and decorate them. In several lineages, however, hair has been lost through evolution, for a variety of reasons. The very largest terrestrial mammals—elephants and rhinoceroses—are nearly hairless because added insulation is superfluous in the tropical environments where they live. Significantly, "woolly" (hair-covered) forms of both these animals were common in northern latitudes during the last ice age. But modern elephants and rhinos have evolved thick, tough skins instead (hence the expression "skin like a rhinoceros hide"). Their large, barrel-shaped bodies retain heat, and a thick fur coat would only prevent them from dissipating excess heat through the surfaces of their bodies. Like the skin of reptiles, the hairless and highly keratinized skin of elephants and rhinos protects them from both water loss and abrasion. Elephants liberally douse themselves with water and mud when they have the chance, in order to cool and protect themselves.

At the other end of the size spectrum is a truly peculiar hairless mammal. The naked mole rat of Africa lives in large social colonies underground. Mole rats are completely subterranean and spend their time burrowing through warm ground in search of tubers to eat. We believe that their nakedness evolved as a means to help them maintain even body temperatures while living at close quarters in a warm underground environment. Like many burrowing mammals, naked mole rats are small, slender, sausage-shaped creatures with large, gnawing incisor teeth, a rather unfortunate combination unless you happen to be an animal of the same species (see color plate 4).

Many mammals have evolved hairlessness as part of adapting to an aquatic lifestyle. The skin of dolphins, for instance, is hairless and is covered with an extraordinary form of slight, fingerlike ridges that produce an even, layered flow of water over the animal's surface, reducing drag as the dolphin swims and allowing it to move more quickly and efficiently.[15] Dolphins, of course, are famous because they are so intelligent and because they cut such dashing figures in the water. But there are many other naked aquatic mammals with less glamorous images. My personal favorite is the hippopotamus.

Hippos are large-bodied terrestrial mammals that spend most of the day in the shallow water of rivers and lakes. They come onto land at night to feed, sometimes walking several kilometers in the sunless cool of the evening to find appropriate grass or other vegetation. The stratum corneum of a hippo's skin is thin, smooth, and dense. This layer permits the animal to lose water from the surface of its body at a rate higher than that of any other mammal we know. This process is called transepidermal water loss. This isn't sweating—hippos actually lack sweat glands—rather, it is direct water loss from the skin's surface, which helps them dissipate the high heat loads they build up both in the water and on land. Hippo skin also contains unique glands that secrete a viscous, pinkish fluid ("red sweat") that helps to protect against the sun (see color plate 5). Even with its natural sunscreen, however, hippo skin dries up and becomes unable to transport water through its surface if the animal is out of the water for very long. It is probably this physiological anomaly that keeps hippos in the water most of the time and ties them to living near permanent water sources. This idea is supported by studies showing that the high rate of water loss through the stratum corneum in hippos helps to keep them cool, much as sweating does in other species. Hippos further reduce their heat loads and reduce the amount of water they lose by evaporation by feeding at night.[16]

I became fascinated with hippos and their skin as a result of doing paleontological fieldwork in Nepal several years ago. There, as well as in northern India and Pakistan, the bones of extinct hippos litter the landscape. I wondered why the animals had become extinct. I realized that the answer to this question was probably connected to their complete dependence on permanent water sources because of the delicate and permeable nature of their skin. The pattern of seasonal monsoon rainfall that is common throughout southern Asia today became intense a few million years ago. As rivers in Asia also became highly seasonal in their patterns of flow through time, hippos would have been among the first mammals to suffer.[17] As rivers began to run dry at certain times of the year, hippos were left literally high and dry. Sadly, their skin was their ticket to oblivion.

FIGURE 11. The evolutionary position of humans among primates is best expressed in a cladogram. Humans and chimpanzees are each other's closest relatives and share a common ancestor that would have looked different from either of them today. Deeper in time, this common ancestor shared a common ancestor with gorillas, and so forth. This entire group of animals—including the other African apes, orangutans, gibbons, Old World monkeys, and humans—constitutes the zoological infraorder called the Catarrhini; these animals are thus referred to as catarrhine primates. (Courtesy of Andrew Lax.)

As intriguing as hippos may be, they are generally not the first animals that most people think of when asked about hairless mammals. That distinction goes to humans, the naked apes. Our bodies are functionally naked and, in this respect, differ from those of even our closest evolutionary relatives. Questions about how and when the human body evolved its nakedness have been asked for years. The best way to gain insight into these issues is to use the evolutionary relationships among ourselves and our close relatives as a guide.[18] Because people are interested in their own history, researchers have studied the evolutionary relationships between people and other animals for the past 150 years, with increasing intensity and sophistication. As a result, the working phylogeny (model of evolutionary history) for the primate group to which we belong is supported by a great deal of evidence. We can visually describe this phylogeny using a cladogram (figure 11).[19]

Our nearest living relatives are common chimpanzees. This relationship does not mean that we evolved from chimps, but rather that chimps and humans shared a common ancestor in the past that looked like neither group does today. One of the most compelling facts to emerge from the past twenty years of molecular studies of primates is that people and chimps are more closely related to each other than either is to the gorilla.[20] Based on appearance alone, most people would guess that chimps and gorillas were the most strongly related pair, but looks can be deceiving. Many of the physical similarities of chimpanzees and gorillas—namely, their hairiness, tooth form, body proportions, and knuckle-walking mode of locomotion—are traits they have carried with them from a common ancestor that existed an estimated eleven million years ago. Chimps and gorillas likely retained these shared features because they have remained mostly in forests of equatorial Africa similar to those that sheltered their common ancestor. In such a relatively stable environment, natural selection would tend to produce few major changes. In terms of appearance, humans really are the "odd man out." Since the human and chimp lineages diverged from their common ancestor, humans have undergone dramatic and relatively rapid anatomical change, resulting in our long-limbed, small-toothed, big-brained, and effectively hairless bodies. These changes are the direct and indirect results of natural selection that produced adaptations to more open environments, such as the woodland savannah, in which most human ancestors are thought to have lived. Interestingly, the anatomical differences between humans and chimps that we see as so dramatic turn out to have been produced by relatively few major genetic changes. As one anthropologist put it, we are 98 percent chimpanzee.[21]

The starting point, then, for understanding the evolution of human skin and skin color is the skin of our close primate cousins. All of these animals have skin that shares three key anatomical features and functional properties.[22] First, it is thicker on the back of the body than on the front, and it is covered with hair, which can provide insulation and protection. In some of our relatives, such as Japanese macaques, this hair is very dense; in others, such as chimpanzees, the hair is much thinner. In humans, the hairs are

FIGURE 12. In this group of chacma baboons *(Papio hamadryas ursinus)*, the infant in the background clinging to its mother's back has a light-colored face. The mother and the other adults have dark faces after years of sun exposure, which has stimulated the production of melanin pigment in their skin. (Photograph courtesy of Mauricio Antón.)

mostly so thin as to be nearly invisible. Second, primate skin can produce sweat. For primates, sweating is essential because it is the primary way the animals stay cool in the heat or during exertion. The amount of sweat that can be produced in the skin varies from one species to another, but all primates are capable of sweating. Third, primate skin can produce the dark pigment melanin. This ability varies by species and according to the conditions in which the animals live, but all primates have melanocytes in their skin that can manufacture melanin.

Although we tend to think of our ape and monkey relatives as being dark and hairy, close inspection of their skin yields some fascinating surprises. With few exceptions, all of these animals are born with pale skin, which remains only lightly pigmented under their hair as adults. The skin that is regularly exposed to sunlight—on the face, hands, and feet—gradually becomes very dark over the course of months and years. This is easily seen in a "family shot" of baboons (figure 12). Similar changes occur in chimpanzees, where the pale complexions of infants contrast dramatically with the darker complexions of older animals. When the young animals are kept indoors (as in laboratories and some zoos), their skin never darkens.

It is likely, then, that the common ancestor of humans and chimpanzees had light skin covered with dark hair, a condition that was the ancestral state for the entire extended lineage that includes monkeys, apes, and humans.[23] From this starting point, we can now look at how humans lost most of their hair. The answer is linked to the reason humans became the sweatiest of all primates and evolved skin of different colors.

3
sweat

Explanations for the nearly hairless state of the human body abound. Without direct fossil evidence that could document the timing and context of hair loss in humans, scientists have proposed likely evolutionary scenarios using comparative anatomical, physiological, and behavioral information, as well as varying amounts of imagination. The resulting hypotheses range from the well-founded to the wacky, with hairlessness being attributed to everything from a heritage of swimming to nit-picking. The best-supported theories involve the importance of sweat in human evolution—the topic of this chapter. It might be useful, however, to take a look at some of the other ideas that have gained notice over the years.

The explanation that has had the greatest popular appeal is the so-called aquatic ape hypothesis.[1] According to this idea, the beginning of the human lineage around six to seven million years ago was marked by an aquatic phase, during which ancient hominids—a term that describes all the members of our lineage since we last shared a common ancestor with chimpanzees—lost most of their body hair, gained a layer of body fat under the skin (subcutaneously), and made the transition from being routinely on four legs to standing and moving on two (that is, became bipedal).[2] As evidence of the aquatic phase, this hypothesis notes that fossil remains of our distant hominid relatives are most often found in the vicinity of ancient lakes. It also observes that several of the anatomical characteristics of modern humans, such as hairless bodies and subcutaneous fat, are shared by aquatic mammals like dolphins and whales. In the peak years of its popularity (the 1970s), the aquatic ape theory—which sounded relatively simple and described our ancestors engaging in pleasant activities like swimming and carrying babies in the water—seemed an attractive alternative to the other hypotheses then available for the emergence of humanity, many of which cast our beginnings in a context of mindless violence and conflict.[3]

The aquatic ape hypothesis is not, however, supported by facts. Let's look at the situation in which our distant ancestors found themselves in tropical Africa. First, a hominid ancestor who spent much of its time in a lake would have had to enter the water from a shoreline. For millions of years, the shores of African rivers, lakes, and waterholes have not been friendly places. Thick with crocodiles constantly on the prowl for hapless prey, shorelines are dangerous places where few animals linger. Our hominid ancestors, only about one meter (a little over three feet) tall and lacking claws, big teeth, or weapons, would have been no match for such formidable predators.

Even if they managed to get into the water, ancient hominids would have faced other big problems. Human skin has few defenses against the waterborne parasites that live in African lakes and rivers. In the African tropics, one of the greatest health risks for people who spend time in and around water is schistosomiasis, a parasitic infestation caused by a tiny worm that can swiftly penetrate and migrate through the skin. Many other parasitic diseases are transmitted to humans this way, leading to untold numbers of deaths and a calamitous loss of vigor and livelihood for many populations who depend on the water.[4] If hominid ancestors had lived in an aquatic habitat during their early evolution, the human immune system would reflect a history of assault by such parasites. It doesn't. Only in the last ten thousand years or so have we started spending much time in the water, as we developed agriculture and fishing, and our immune systems have not yet been sharpened by natural selection to resist the attack of the nasty organisms that inhabit these freshwater lakes and rivers.

The aquatic ape hypothesis also doesn't adequately explain why hominids would have evolved hairless skin. Naked skin is advantageous to fully aquatic animals like whales and dolphins because it reduces drag and buoyancy. This is especially important when the animal is diving or speeding along in the water, either in search of food or on a long-distance migration. But we have no evidence to suggest that hominids ever engaged in such activities, although they may have occasionally foraged along shorelines and in shallow water for shellfish and similar food.[5] For animals who spend only part of their time in the water, a naked skin is in fact something of a liability because it leaves them vulnerable to problems of thermoregulation when they are on land. Animals like otters, fur seals, and sea lions, which weigh less than 1,000 kilograms (or less than a ton) and spend some time on land, are covered with smooth, dense fur that insulates them against the cold when they are out of the water. Only the giant semi-aquatic mammals such as walruses and hippos, whose large size and barrel shape make it hard for them to lose heat from the body's surface, have naked skin.

In addition, all mammals committed to a fully aquatic or semi-aquatic lifestyle have evolved a streamlined body shape with small appendages such as fins or flippers to improve their hydrodynamics and minimize the area of skin in contact with the water.[6] If you spend much of your life in the water surrounded by watchful and hungry predators, you need to be able to move fast and maneuver deftly in the water. The ancient hominids, in contrast, were fairly short, bipedal apes, with long gangly arms, who wouldn't have been able to dog-paddle or defend themselves in the water for five minutes, let alone spend their days diving and cavorting while swimming after prey or one another.

In short, we didn't lose our body hair because we passed through an aquatic phase. All the human features that this hypothesis ascribes to a heritage of swimming and diving can be explained far more convincingly and parsimoniously as our adaptations to an active life in a hot and mostly open environment.[7]

Other theories about the evolution of human hairlessness have also garnered popular attention. One recent hypothesis argues that hairless skin is advantageous to survival and reproduction because it reduces the disease toll caused by parasites that infest hair and fur.[8] Because fur and feathers are warm and inviting traps for disease-causing ectoparasites such as lice and ticks, a hairless body makes it much harder for such creatures to establish an infestation. It became possible for hairlessness to evolve, according to this hypothesis, when humans adopted clothing and more effective ways to shelter themselves from the elements. Why maintain a permanent layer of fur when you can simply take off your clothes and wash them when they become infested with parasites?

This theory assumes that humans must have had clothing and the means to build insulated shelters before hairlessness could evolve. But there is no evidence that this was the case—and in fact we have abundant evidence to the contrary. Clothing and shelter are fairly recent inventions in human history. Artifacts such as awls and needles—indirect evidence of clothing, as they could have been used to sew animal skins into simple garments—are confined to the last forty thousand years of human history and are found primarily outside the tropics.[9] The ectoparasite hypothesis asserts that even relatively early hominids such as *Homo erectus* in Africa had the cultural means (clothes, shelter, and fire) to allow hairlessness. The archaeological record, however, offers no evidence to support these assertions. Further, nothing in the historical ethnographic record indicates that people pursuing traditional lives in the tropics of Africa, Australia, or elsewhere ever wore very many clothes, even under cold conditions. Today, the indigenous peoples of the tropics are generally the most hairless of all humans (figure 13). Hairlessness was almost certainly the original or ancestral state for modern humans, and its origins had nothing to do with reducing ectoparasite loads by wearing clothes.[10]

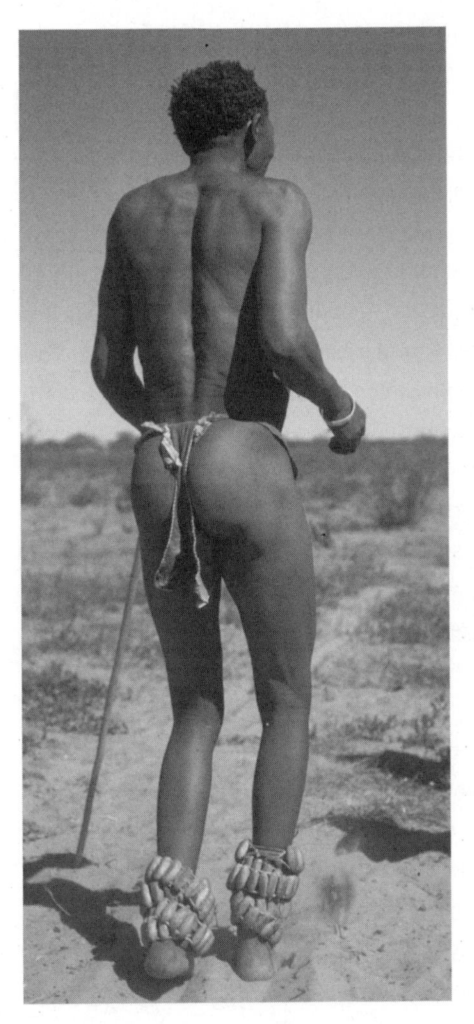

FIGURE 13. Most of the indigenous peoples of the tropics, such as this San (Basarwa) tribesman from Botswana, entirely lack visible body hair and traditionally wore few or no clothes. (Courtesy of Edward S. Ross.)

The only explanation for the evolution of hairlessness that is consistent with available fossil, anatomical, and environmental evidence centers on the importance of sweat. For an active primate living in a hot environment, having a functionally naked and actively sweating skin is the best way to maintain a steady body temperature and—literally—to keep a cool head. But we must first ask why humans are naked when other animals (including other primates) living in the same environment are not. What happens to animals with thick coats of fur when they are active in hot environments?

In the heat caused by strong sunlight, a layer of fur or feathers reduces the amount of heat that an animal gains from its environment. Although it sounds counterintuitive, for most animals, having a heavy coat actually keeps them cooler in the sun because their coat traps heat (absorbs short-wave radiation) and then emits it (as long-wave radiation) back into the environment before the skin itself has a chance to become significantly hotter.[11] This works fine when the coat is dry, but it's a problem when the coat becomes wet, as it would be if moistened by sweating.

When an animal generates more heat through its own activity or when the temperature goes up outside, the animal must cope with a rising internal heat load. Many mammals do this by sweating. The evaporation of sweat cools an animal because heat is lost from an object when liquid vaporizes from its surface. The most efficient evaporative cooling afforded by sweating occurs right at the surface of the skin itself. But if an animal builds up a sweat and its coat gets wet, most of the evaporation will occur at the surface of the coat and not at the surface of the skin. This leads to a buildup of heat in the body because heat from the blood vessels in the skin must be transferred to the surface of the wet fur rather than being dissipated quickly from the skin's surface. As a result, the animal ends up sweating much more, and its fur gets wetter in order to achieve an adequate degree of cooling. Physiologically, this is highly inefficient and nearly impossible to sustain for any length of time. Unless the animal can drink regularly while it exerts itself, it collapses of heat exhaustion.

Sweating becomes more important as environmental temperatures rise or as an animal engages in more strenuous activity. When animals or people are at rest or engaged in low to moderate levels of physical activity, they stay cool by means of radiation (a transfer of heat from one object to a cooler one), convection (a heat exchange by physical transfer, as in the case of air currents), conduction (a heat flow from one object to another by direct contact), and evaporation (figure 14). When outside temperatures are hotter,

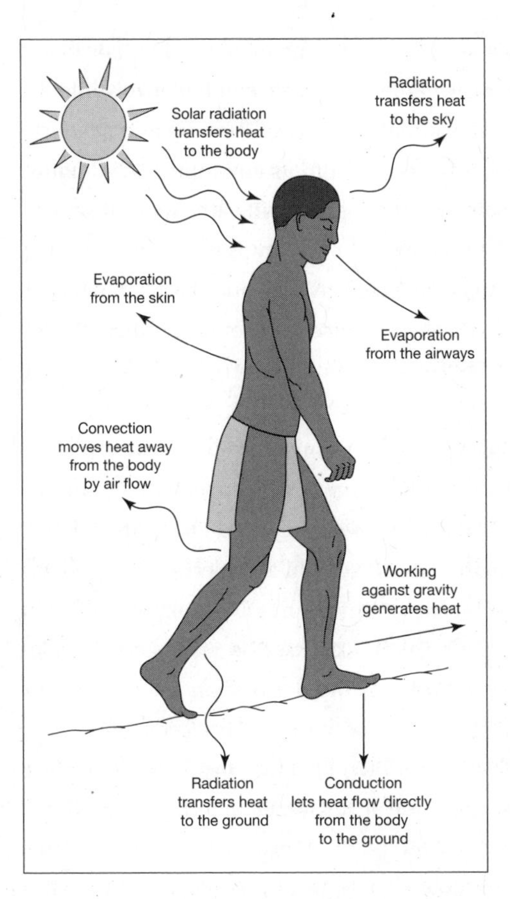

FIGURE 14. People and animals stay cool by means of radiation, convection, conduction, and evaporation. Evaporative heat loss through sweating becomes more important as environmental temperatures rise or as activities become more strenuous. (Illustration by Jennifer Kane.)

the differential between body and air temperatures is lower, which restricts the amount of heat that can be lost by means other than sweating. As an animal's activity level increases, it produces more heat—especially in the large muscles—as the result of elevated metabolism. This heat has to be dissipated before it reaches a dangerous level. The problem is compounded when strenuous exertion is combined with high air temperature. Under these conditions, the body's ability to lose heat efficiently by evaporation is essential for survival, and anything that slows or prevents this process endangers the animal's life. A hominid with a thick coat of hair would have a hard time keeping cool when it was highly active because its wet hair would act as a blanket, impeding the loss of heat from the skin's surface. The body's efforts to produce more sweat in a vain attempt to keep cool would then result in rapid fluid loss. Most authorities now agree that these are precisely the conditions that triggered the evolution of hair loss in the human lineage. Remove most of the body hair and the problem of evaporating sweat from the surface of the skin disappears.[12]

Human skin contains mostly eccrine sweat glands, which secrete copious amounts of a thin, watery fluid that evaporates quickly during the heat-reducing process technically referred to as thermal sweating.[13] In contrast, other mammals possess greater numbers of apocrine sweat glands. These glands produce small amounts of a milky, viscous fluid that dries in glistening, gluey droplets. In animals such as horses, apocrine sweat produced during exercise mixes with sebum from the sebaceous glands and forms a lather that helps the animal cool off. (This is the origin of the expression "in a lather.") Human adults have relatively few apocrine glands, most of which are found in the groin, armpits, and external ear, where they produce secretions in response to stress and sexual stimulation. In human evolution, eccrine glands came to greatly outnumber apocrine glands for a very good reason.

Animals with heavy fur coats and mostly apocrine sweat glands can produce only 10 to 20 percent as much sweat as a competent human "sweater" under conditions of high heat or strenuous exercise. For those animals, sweating alone does not cool the body sufficiently to keep their sensitive organs within tolerable temperature limits. Consequently, many of these animals have evolved other mechanisms such as panting to keep their delicate organs cool (figure 15).[14] Most mammals other than primates use a combination of apocrine sweating and other means, including panting. Nonhuman primates keep cool only through sweating, with the sweat itself produced by varying mixtures of eccrine and apocrine sweat glands.[15] Many species also rely on behavior for temperature regulation: they simply rest or seek shade when it is too hot, to prevent the buildup of body heat. Why, then, did humans in particular evolve into such proficient eccrine sweat machines?

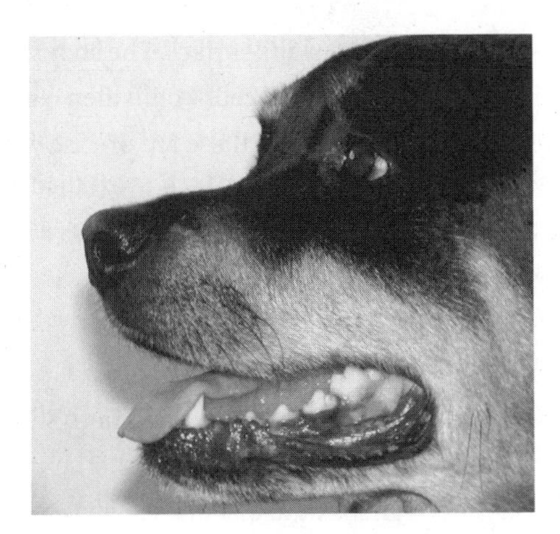

FIGURE 15. Dogs and other carnivores keep cool by a combination of apocrine sweating and panting. The evaporation that occurs within the mouth during panting cools the blood flowing through the veins there. This cooled blood then tracks back deep into the skull, where it helps to reduce the temperature of the base of the brain before returning to the heart. A dog also loses some body heat by evaporation from the inner surfaces of its nose.

The emergence of the genus *Homo* around two million years ago was a turning point in human evolution in many ways.[16] As these hominids evolved, eccrine-based thermal sweating became increasingly important to them for two primary reasons, both related to key features of their development. The first was the rise in their activity levels, especially during daylight hours when most other animals must retreat to the shade. The second was a significant increase in average brain size.[17] In both these cases, more efficient cooling of the body was crucial.

Early members of the genus *Homo,* exemplified by the lanky juvenile "Turkana Boy," who lived on the shores of Lake Turkana in northwestern Kenya 1.6 million years ago, were taller and more long-legged than their ancestors, with relatively shorter arms.[18] Their brains were larger than those of their predecessors, and they were also becoming more modern in their lifestyle, engaging in vigorous activity such as long-distance walking in hot, open environments, rather than being routinely bound to the safe refuge of forests. Evidence from many sources—studies of the remains of hominid bones and teeth, preserved stone tools, and other remnants of this ancient material culture, as well as the modern physiology laboratory—tells us that these ancestors were active, striding bipeds who ate eclectic diets and walked great distances in search of materials for making tools.[19] Their increased range of movement was especially notable: recent studies comparing the anatomy of some ancient *Homo* fossil skeletons and modern humans during exercise have demonstrated that long-distance running may have been the activity that influenced the shape of the modern human body more than any other.[20] Adaptations to greater mobility on the open ground, in connection with the pursuit of game, bouts of long-distance foraging, searches for raw materials for stone tools, and other activities essential to survival, were being promoted by natural selection.

Coping with heat was one of the greatest environmental challenges faced by these increasingly active early hominids, and there is no doubt that improved sweating ability was an essential aspect of meeting that challenge, primarily through developing a preponderance of the extremely efficient eccrine glands in the body. On a hot day in the desert, these glands allow most humans to produce over twelve liters (more than twelve quarts) of eccrine sweat, at a rate of about one liter per hour, which can temporarily increase to a prodigious three and a half liters per hour in some people. When this fluid evaporates, the resulting reduction in body temperature makes for a powerful whole-body cooling mechanism. Blood flowing just below the sweating skin is cooled by evaporation and heat loss from the skin's surface. The cooled blood in the veins is then returned to the core of the body to be reoxygenated by the lungs and then redistributed by the heart and arteries to serve temperature-sensitive organs such as the brain, the liver, and skeletal muscles. You can see evidence of this process in your own body. When you build up body heat either from the air temperature or from exercising, the veins in your hands, feet, head, and neck dilate—that is, they enlarge or widen—and become much more visible. In people with lightly pigmented skin, vigorous exercise in the heat will almost always produce a characteristic red face because of the expansion and filling of the surface blood vessels there.

Maintaining an even body temperature is essential for the normal functioning of all the body's internal organs, but this is especially important for the brain. Internal organs cannot operate well at temperatures outside a very narrow range within a few degrees of normal (that is, 98.6°F or 37°C). For example, your abilities to think, reason, and communicate are compromised when the brain is even slightly overheated, as anyone who has ever experienced a high fever knows. Delirium sets in at about 103°F (40°C), and unconsciousness, followed quickly by death, occurs when the brain's temperature remains in excess of 106°F (42°C).

Brain cooling in humans is not a trivial matter because our brains are so much bigger than those of even our large-brained primate cousins, chimpanzees. Our brains occupy an average volume of 1300–1400 cubic centimeters (approximately 79–85 cubic inches), whereas chimp brains average only 450 cubic centimeters (a little over 27 cubic inches). For the last two million years of evolution, the size of the human brain has expanded significantly, to the point where we now have the largest brain of all animals relative to body size.[21] Big brains provide many important services to us, but they are costly to maintain. Like a powerful engine in a racing car, our brains require lots of high-quality fuel, causing them to be extremely sensitive to overheating. Because the temperature of the brain closely follows the temperature of the blood in the body's arteries, the temperature of circulating blood must be carefully regulated. With increased brain size, the importance of having a hard-working and efficient whole-body cooling system comes into play more than ever before.[22] Behind every large human brain, there is a potentially very sweaty human body.

Early members of the genus *Homo* were vigorous, active bipeds. (Interestingly, their bipedalism also helped to reduce their bodies' heat load somewhat. A bipedal primate under the noonday sun at the equator actually has a smaller surface area of skin exposed to direct solar radiation than a quadruped does, which makes it easier for modern humans to be more active than other animals during the heat of the day.)[23] Available physiological, paleontological, and paleoenvironmental evidence indicates that it was at this stage in our evolution that we became highly competent sweaters. The apocrine component to sweating was probably effectively lost by this time, and humans went on to become the world's most efficient eccrine sweat producers.

We can deduce that it was also at this point in evolution that humans lost most, but not all, of their hair. What remained was a strategic mass of thick hair on the top of the head. This may at first seem counterintuitive in light of what we know about the importance of keeping the brain cool, but in fact maintaining a thick head of hair was critical in human evolution. First, it protected the scalp from the damaging rays of the sun, preventing sunlight from beaming directly onto the surface of the scalp. Second, it facilitated brain cooling because when the sun was high, the surface of the hair heated up, leaving a barrier layer of slightly cooler air next to the scalp. The scalp can then lose heat efficiently into the barrier layer through radiation and evaporation. You can easily verify this the next time you're out in the sun. The surface of your hair will get very hot (especially if your hair is dark-colored), but the air next to your scalp will feel a little cooler. This effect works best with frizzy and somewhat interwoven hair, which creates a thick barrier layer of cooler air between the scalp and the top surface of the hair facing the sun.

In modern humans, eccrine sweat glands are widely distributed over the surface of the body. Tubular in form, they lie in the outer portion of the dermis and—unlike apocrine glands—are not associated with hair follicles (refer back to figure 1). In most mammals, eccrine glands are confined to the friction surfaces of the animals' palms and the soles of their feet, where they help to keep the skin pliable so that the animals can maintain secure footing. The watery fluid produced by an eccrine sweat gland is expressed to the surface of the skin through an individual pore. Humans have two to four million eccrine glands on the body's surface, at an average density of 150–340 per square centimeter. These glands are most numerous on our palms and soles, undoubtedly a pattern that we retain from our mammalian forebears.[24]

Both eccrine and apocrine sweat glands produce sweat in response to heat, and both, to varying degrees, contribute to the process of thermal sweating. This is accomplished by heat-induced stimulation of sympathetic nerves that are part of the autonomic nervous system. The autonomic nervous system in general is concerned with maintaining "housekeeping," or automatic, functions in the body that are not under conscious control: regulating our heartbeat, controlling the diameter of blood vessels, and modulating the size of pupils in our eyes, for example. In particular, the sympathetic division of this system helps to mount the body's reactions to aversive stress in "fight or flight" responses. The eccrine glands of the palms and soles differ from those located elsewhere and respond only to emotional stimuli, whereas those on the face and in the armpits respond to both thermal and emotional stress.[25]

In humans, apocrine glands do not play a large role in cooling the body through thermal sweating, and they are generally considered evolutionary relics. They do, however, serve a thermoregulatory function in many mammals, such as ungulates, that did not evolve eccrine glands for dissipating heat. In our closest primate relatives, chimpanzees and gorillas, eccrine glands are more common than apocrine over the surface of the body, but they are not as dense and plentiful in these animals as they are in humans.

Sweat glands and their activities are of great interest to scientists and clinicians, who have devoted much research to comparisons of the quantity, structure, and function of sweat glands in various groups of people. Oddly, only a few rigorously conducted and strictly controlled studies have compared the thermal sweating responses of individual people. The density of sweat glands on the body actually varies little from person to person and place to place, although small variations are sometimes found between different populations.[26] Occasionally, you will hear a person say, "I never sweat," or "I just can't seem to stop sweating." Such differences in sweating activity appear to have two major causes. First, people differ in the relative number of active versus nonactive sweat glands. Age, weight, sex, and other factors affect the number of active sweat glands in the body. Second, a person's sweating ability depends on how hydrated that person is and how physiologically acclimated to a particular climate he or she has become. These factors may account for the widely reported higher sweating rates of light-skinned people of European ancestry who live in hot places versus those of more dark-skinned Africans and Asians.

Sweating is critically important in regulating our body temperature, but humans use other means, too. Physiologists classify temperature regulation in people as either involuntary or voluntary. Involuntary regulation by the skin is complex and requires a concatenation of reactions, not just sweating. It begins when information about body temperature is transmitted to the brain. If body temperature is outside the normal range, the body's thermostat kicks in by regulating heat transfer between the body's core and the skin. This is done by modifying the diameter of the peripheral blood vessels, which changes the amount of blood circulating through the skin. The skin is the interface for losing heat to the environment or gaining heat from it and the surface that allows the spreading of sweat necessary for evaporative cooling. People are also very good at regulating their temperature by voluntary means, taking many conscious actions to keep body temperature comfortable: seeking shade on a hot and sunny day, putting on or taking off clothes, and using heaters or fans. These activities have become more complex and sophisticated in the course of human history, thus reducing the selective pressure on the human body to adapt itself to environmental extremes.

The relative importance of sweating in body cooling depends on the interaction of the ambient temperature, the humidity, and body activity. Under conditions of extreme heat and humidity—for example, when the outside temperature is greater than our body temperature and the relative humidity exceeds 90 percent—a person cannot dissipate heat without sweating because the body is actually gaining heat from the environment rather than losing heat to it. In such an extreme situation, sweating accounts for 90 percent of the body's cooling ability. The person must continually take in fluid in order to maintain blood volume and keep the sweat glands supplied with coolant.[27] When people exercise vigorously in the heat, their sweat glands must be in top form and perform at their peak level, which requires fluid intake.

Sometimes people sweat for reasons having nothing to do with outside heat or levels of activity. Women normally experience hot flashes and night sweats during the course of menopause, and people suffering from illnesses such as influenza and malaria alternate between bouts of chills and copious sweating. In these cases, the body produces sweat because the neural structures responsible for temperature regulation in the brain or spinal cord have been activated by changes in hormone levels or by fever-inducing chemicals released by invading microorganisms.

Some individuals who suffer from a genetic disorder called anhidrotic ectodermal dysplasia, which causes them to have few or no active sweat glands, have great trouble tolerating strenuous physical activity, especially in hot environments.[28] Even people with a normal complement of active sweat glands need to maintain those glands in good condition to carry out a normal range of daily activities. The ability of sweat glands to respond to heat stress is adversely affected by sunburn, for example.[29] Consequently, protecting sweat glands against damage caused by strong sunlight has been crucial during the long course of human development in the tropics. This is one reason why darkened skin pigmentation is important to the peoples of the tropics today and why it was a key innovation in our past.

Experimental studies and simulations of temperature regulation under the stressful environmental conditions of the hot tropics show that heat loss is most efficient in lean people who have a high ratio of skin surface area to body weight—in other words, heat loss is maximized if the skin's surface area is large and the person is thin.[30] Thus, in very hot environments, tall, lean people have an advantage over short, stocky ones in how well they can maintain thermal equilibrium. This is why many long-term inhabitants of the Old World tropics, such as Nilotic tribespeople, Australian Aborigines, and many of the tribal people of India, are tall and lean, with long, thin limbs (figure 16). Their long limbs provide more surface area for

FIGURE 16. The indigenous peoples of the tropics, such as this native of Pharasgaon, India, often have long limbs and lean bodies. Their increased surface-to-volume ratio helps them to lose heat quickly under hot conditions. This relationship is even more important when evaporation from thermal sweating becomes the dominant mode of body cooling, as when the humidity is high or when a person is exercising. (Courtesy of Edward S. Ross.)

the transfer of heat to the environment, and their thinness speeds the transfer of heat from the body core to the surface.

This relationship is the basis for Allen's Rule, which states that mammals living in cold regions will minimize the size and surface area of their extremities, whereas those inhabiting hot areas will increase the relative size of appendages. Examples of this rule are abundant in nature. Whether we look at ground squirrels, mice, rabbits, or any number of other mammals, those that live in cold or high mountainous areas have shorter feet, ears, and tails than those that live in warmer lowlands or deserts. The same principles help us to understand why overweight or obese people are often under heat strain and sweat a lot. As the thickness of the body's skin and subcutaneous fat layer increases, the rate of heat flow from the body's core to the exterior slows, and there is less opportunity to lose heat to the environment through radiation, convection, or evaporation.

The humble sweat gland thus must assume pride of place in human evolution. Without plentiful sweat glands keeping us cool with copious sweat, we would still be clad in the thick hair of our ancestors, living largely apelike lives. We would never have evolved our hot-running, high-octane brains or our ability to be active and alert even in the heat of the day in the hottest places. No one has composed paeans or odes to celebrate the glories of sweat, but well they should. It is plain old unglamorous sweat that has made humans what they are today.

4
skin and sun

As we go about our daily lives, our skin is always active, and its complex chemistry is constantly changing. Skin cells are dividing, important molecules are being broken down, others are being repaired, and yet others are being created on the spot. Because the human lineage originated in the tropics and spent most of its six million or so years of existence in tropical areas, part of the skin's activity has involved a series of anatomical and biochemical adaptations to heat and sunlight. Sweating is only part of the story. Our skin has also evolved other ways to mediate vital chemical transactions between the body and the environment, and particularly between the body and sunlight.

The sun emits a wide variety of electromagnetic radiation, ranging from very short-wavelength ionizing radiation such as gamma rays to very long-wavelength infrared radiation and radio waves (figure 17). Ultraviolet radiation (UVR) itself includes a broad spectrum of wavelengths, from very short-wavelength vacuum UV to longer-wavelength UVC, UVB, and UVA. Although it is maligned by most biologists because of its destructive effects on bio-

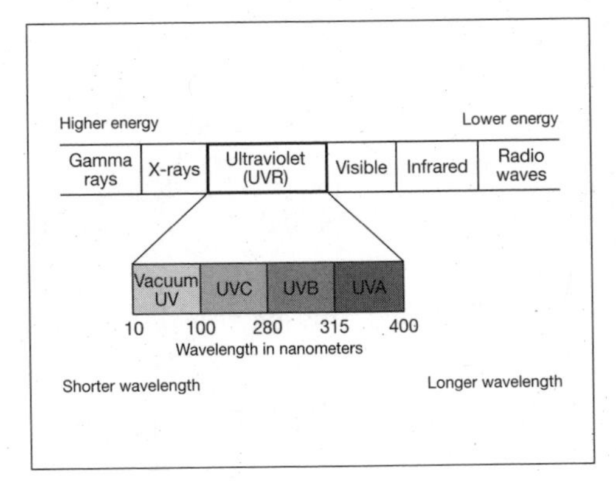

FIGURE 17. Solar radiation comes in a wide range of wavelengths and energy levels. The most harmful types of radiation are those with the shortest wavelengths and the highest energy, such as gamma rays and UVC. Oxygen and ozone in the earth's atmosphere filter out much of the harmful UVR, but overexposure to UVR can damage DNA and destroy folate in the body. (© 2005 Jennifer Kane.)

logical systems, UVR has been one of the most important forces in the evolution of life on earth. From the early days of our planet's history, unicellular and multicellular organisms have been forced to evolve mechanisms to protect their delicate chemical reactions from its disruptive effects.[1]

For living creatures, the most damaging types of radiation are those with the shortest wavelengths and the highest energy, such as gamma rays and UVC. During the long sweep of earth's history, our planet has developed an atmosphere rich in oxygen and ozone, which very effectively screens out the most harmful wavelengths of solar radiation. Longer wavelengths of UVR (specifically UVB and UVA), along with visible light and infrared and radio waves, penetrate the atmosphere much more easily. Scientists and some politicians today are rightly concerned about the health of the earth's atmosphere and, in particular, the state of our protective ozone layer. If this ozone layer becomes thinner or badly perforated, most forms of life on earth would suffer from the destructive effects of high-energy solar radiation, especially excessive UVB.

Using satellite data, we can generate a map showing average UVR levels at the earth's surface (see color map 1).[2] This map reveals some predictable patterns and a few surprises. Levels of UVR are highest in the tropics and especially at the equator. But latitude is not the only determinant; some places near the equator experience much higher levels of UVR than others.

Arid regions like the Sahara Desert receive very high levels of UVR, whereas more humid or cloudy equatorial areas like the Amazon rain forest receive lower levels. Outside the tropics, UVR levels are generally lower, with a few conspicuous exceptions. The Tibetan plateau, for instance, experiences very high levels of UVR because of its high altitude and thin atmosphere.[3]

Different types of UVR penetrate the earth's atmosphere to different extents. As UVR approaches the earth, the most high-energy wavelengths (UVC and about 90 percent of UVB) are absorbed by the oxygen and ozone in the atmosphere. The remaining 10 percent of the UVB and all of the UVA pass through, but the amount that actually makes it to any given spot on the earth depends on the latitude and the angle of the sunlight at that particular place and time. As one moves away from the equator and the angle of the sunlight decreases, the atmosphere becomes thicker and filters out more UVB. Thus, low levels of UVB fall on areas at high latitudes. Very small changes in UVB levels have substantial effects on plants and animals. Organisms living at extremely high northern and southern latitudes, for example, have adapted to receiving only tiny doses of UVB, and only at the peak of summer.[4] The unequal distribution of UVB and UVA on earth has had enormous consequences for the evolution of life at different latitudes, and for the evolution of human skin color, as we will see later.

Most of the chemical reactions that UVR causes in the body are harmful. If you've ever had a sunburn, you know that your skin has been damaged—you can feel and see it. But a sunburn is only the most immediately palpable and visible negative effect of UVR exposure. The most serious destruction wrought by UVR is more sinister because it can go unnoticed for years. At the molecular level, UVR can damage DNA, the most important information-carrying molecule in the body, which is essential to cell division. UVR can affect a DNA molecule directly by changing the molecule's chemical composition when it absorbs the UVR or indirectly through the potentially destructive free radicals generated by UVR.

The worst damage is caused when DNA absorbs shorter-wavelength

UVR (mostly UVB). Specific chemicals, known as "photoproducts" because they are the result of solar radiation, are then produced within the DNA molecule, causing small physical distortions in its structure. These distortions are normally corrected by a process known as nucleotide excision repair, which removes and replaces the damaged DNA strands—a sort of corrective surgery at the molecular level. The evolution of the ability to repair DNA was one of the most important innovations in the history of life on earth. The repair usually proceeds uneventfully, provided that not too much DNA was affected and that the repair mechanism itself is in good shape. But if the repair is inadequate, cells will then reproduce with faulty DNA. Given enough time and continued exposure to UVB, cells with faulty DNA can build up in the skin, eventually leading to skin cancer.[5] UVA also causes considerable harm to DNA, although this damage is somewhat different in its structure, and possibly in its effect, from that caused by UVB. UVA has been implicated as a major culprit in the premature aging of skin caused by sun exposure (known as photoaging), and it has been associated in epidemiological studies with the most dangerous form of skin cancer, malignant melanoma.[6]

DNA is not the only molecule adversely affected by UVR. Folate, for example, is a water-soluble B vitamin that is necessary to produce DNA. The body must constantly manufacture new DNA, because so many of its routine functions require cell division: creating new blood cells; replenishing the skin, hair follicles, and linings of the mouth and gut; and producing sperm cells in men (a process that continues throughout adult life).[7] But a lack of folate will slow or curtail DNA production, and all the processes that require new DNA will suffer, especially those that need the DNA quickly or continually. DNA is especially critical for fueling the rapid cell division that occurs in a developing human embryo or fetus, particularly in the early weeks of pregnancy, when organs are beginning to form and the overall plan of the body is being laid out. Without sufficient folate in a mother's body, not enough DNA will be produced to promote the cell division that allows embryonic tissues to differentiate and grow.[8]

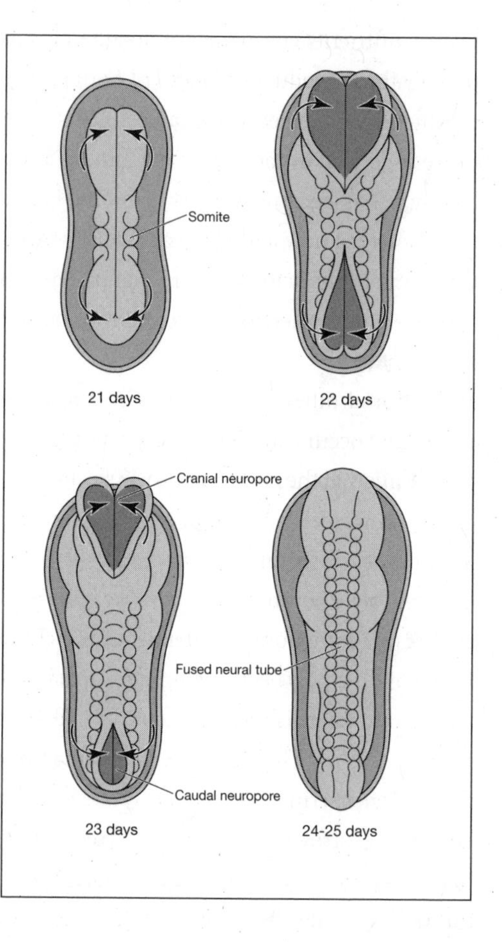

FIGURE 18. The nervous system in a human embryo begins to develop with the formation of the neural tube during the first few weeks of pregnancy. This process involves the precise, zipperlike closing of neural folds on the top of the embryo, as depicted in the sequence here. For this process to succeed, cell division in the neural folds must occur exactly on schedule. If there is not enough folate to fuel DNA production for the rapidly dividing cells of the neural folds, a neural tube defect can develop. The somites, structures flanking the neural tube, are the building blocks for many of the body's future muscles and bones. (Illustration by Jennifer Kane.)

Folate also plays a crucial role in the development of the embryonic nervous system. A shortage of folate at a critical time in early development can lead to birth defects of varying degrees of severity, including some that may be lethal. Folate deficiency is now widely acknowledged as a risk factor for numerous complications of pregnancy as well as for a group of birth defects called neural tube defects (figure 18).[9] The neural tube is the forerunner of the central nervous system in the embryo. It extends from the top of the primordial brain (the cranial neuropore) to the end of the spinal cord (the caudal neuropore).

Few nutrients compare with folate in terms of its impact on health, especially the reproductive health of both men and women. We derive folate mostly from green leafy vegetables, citrus fruits, and whole grains. Because folate is so vital to the body's machinery and to maintaining reproductive health, it has become a focus of public health campaigns in many countries. Folate is now added to many foods (especially breads and cereals) in the form of folic acid to ensure that people sustain adequate levels in their bodies, and women of reproductive age are encouraged to take folate supplements.[10]

UVR and other high-energy radiation can destroy folate in the body. When this occurs suddenly or on a large scale, the consequences are serious because all the chemical processes that depend on folate are affected.[11] Although scientists first documented the adverse effects of UVR on folate in humans nearly thirty years ago,[12] only in the past decade, as the importance of folate became clear, have we begun to understand the implications of this phenomenon. The details of the chemical destruction of folate by UVR have recently been documented in laboratory experiments, which have shown that folate is most susceptible to damage by the longer wavelengths of UVR, UVA.[13] These studies are paving the way for investigations of how naturally occurring UVA affects folate levels in humans during real life. If UVR can destroy folate, a substance essential to human life and reproduction, it is clear that some sort of defensive mechanisms must have evolved through natural selection to help maintain folate levels in the body. This is where the story of skin color comes in, as chapters 5 and 6 explain in greater detail. When the body is suffering from the effects of sun, the evolutionary solution has been to add natural sunscreen to the body's surface, the skin.[14]

Despite the destructive effects of UVR, it is not universally harmful and in fact has some positive biological effects. The most important of these—the production of vitamin D, popularly known as the "sunshine vitamin"—occurs in the skin.[15] Vitamin D exists in several forms: vitamin D_3 is made by vertebrates, and vitamin D_2 is the primary form found in plants. Vitamin D is a unique natural molecule that first appeared on earth as a product of photosynthesis in tiny marine phytoplankton over 750 million years ago.[16] Although its function in the earliest vertebrates (ancient fish) is not well understood, by about 350 million years ago—the time the first tetrapods emerged, the first animals to spend significant amounts of time on dry land—this vitamin had taken on an essential role in vertebrate evolution.

Vitamin D is important to all vertebrates because it allows them to absorb calcium from their diet and build a strong internal skeleton. Fish can easily get enough vitamin D by eating plankton or other fish that contain it. For the earliest land-living vertebrates, however, these sources of vitamin D were not available, even though their need for calcium in order to maintain a rigid skeleton was great. At this stage in evolution, with natural selection operating full force, vertebrates developed the ability to make their own vitamin D. Because vitamin D is manufactured by a photochemical, or sunlight-induced, process, early tetrapods could help to satisfy their body's requirements for the substance by exposing themselves to sunlight. In this way, they could potentially get vitamin D both from their diet and from the vitamin factory in their skin.

UVR in the UVB range stimulates the production of vitamin D_3 in the skin. High-energy UVB photons first penetrate the skin and are absorbed by a cholesterol-like molecule residing in the cells of the epidermis and dermis, which catalyzes the formation of a molecule called previtamin D_3. This precursor molecule is then transformed in the skin at body temperature to vitamin D_3, which undergoes further chemical conversion in the liver and kidney to become the biologically active form of the vitamin. This reaction is self-limiting: if the body's circulation already contains enough of the active form of vitamin D, the process of making more is discontinued, and the chemical precursors are broken down into various inert by-products. In this way, the body averts "vitamin D intoxication," or vitamin D poisoning, the overproduction of the active form.[17]

The active form of vitamin D is used throughout the body for a variety of purposes. It regulates calcium and phosphorus metabolism, the basis of making a strong skeleton. It also facilitates calcium absorption from the

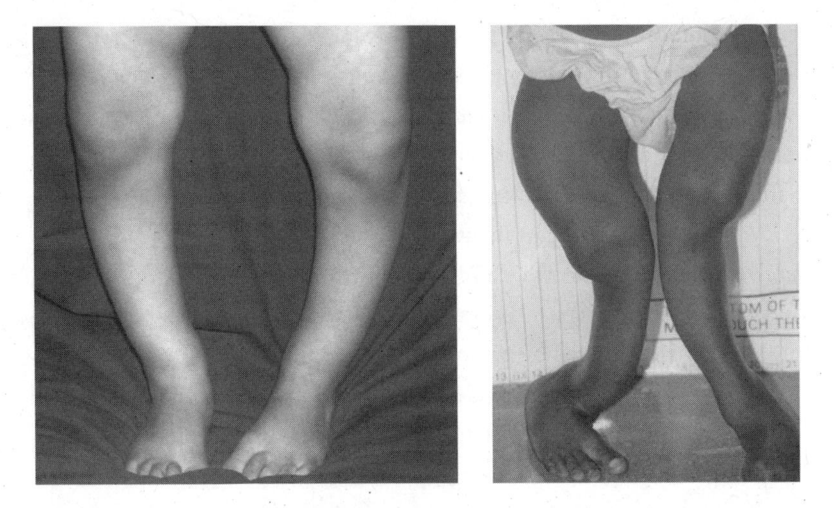

FIGURE 19. The soft, poorly calcified bones of children afflicted with nutritional rickets are bent under the weight of the body. This condition is caused by a serious vitamin D deficiency, which prevents absorption of calcium from the diet. Historically, rickets afflicted mostly children living in far northern latitudes with little UVB, but it is becoming increasingly common in children with dark skin at all latitudes who lack calcium in their diet or who are not exposed to sunlight. (Left, © NMSB/Custom Medical Stock Photo; right, courtesy of Tom D. Thacher, MD.)

gut and has a direct effect on bone-forming cells. We have long known that vitamin D is necessary for the growth of bones because it allows the body to absorb calcium from food.

A shortage or lack of vitamin D has negative effects on the body throughout the human lifespan. A vitamin D deficiency during childhood or adolescence can reduce a person's reproductive ability later on. The most serious and infamous condition caused by vitamin D deficiency is nutritional rickets, a childhood disease in which the long bones of the legs bow under the body's weight (figure 19). In children with rickets, the cartilage in developing bones fails to mineralize properly because the body is not absorbing calcium and phosphate. Serious cases of rickets in girls also prevent the pelvis from forming normally, which can cause later problems with pregnancy, including obstructed labor and a high incidence of infant and maternal health problems and mortality. Abnormally low levels of vitamin D can also interfere with normal ovarian function. Among pregnant women, vitamin D deficiency contributes to pathologically low calcium levels in their blood and later to rickets in their babies. Among all adults, it can produce osteomalacia, a painful softening of the bone's structural framework, and can also affect the functioning of the immune system.[18]

Less widely known than vitamin D's importance to a healthy skeleton is its role in regulating normal cell growth and inhibiting cancer cell growth.[19] Insufficient vitamin D has recently been related to an increased risk of several types of cancer that commonly afflict people in industrialized countries, namely, colon, breast, prostate, and ovarian cancer.[20] These cancers appear to be particularly prevalent among people with chronic, low-level vitamin D deficiencies who live in high latitudes—a finding whose evolutionary significance will become evident in the following two chapters.

Ultraviolet radiation has been a relentless force in the evolution of life on earth. Because of its destructive power, organisms have had to evolve sophisticated means to protect their most basic reproductive machinery—DNA and its folate precursors—from annihilation. Like most villains, however, UVR has a good side that isn't often taken into account. Its ability to transform molecules in the skin into the precursors of vitamin D has been of paramount importance to all vertebrate organisms living on land, including people. The real trick in evolution has been to figure out a way to control the amount of UVR entering the skin, and that is the skin's dark secret.

5
skin's dark secret

Human skin is inherently colorful. Within our single, recently evolved species, *Homo sapiens,* skin colors make up an exquisite palette, varying in almost imperceptible degrees from the palest ivories to the darkest browns. This array exists because people differ in the amount of melanin pigment their skin contains and the ways in which it is packaged. Melanin, from which human skin derives most of its pigmentation, is a remarkable molecule that has had literally thousands of uses in the evolution of life. Its role in protecting human skin is only one of its more recent.

Melanin is the name given to a family of complex polymeric pigments that exist in many forms. (A polymer is a chemical compound composed of multiple repeating units.) The primary form of melanin we find in the human body is an extremely dense, almost insoluble, and very dark brown pigment molecule that is attached to a protein.[1] When this form of melanin is isolated in the laboratory, it looks like sludge in the bottom of a beaker. Melanin pigments are widespread in nature, imparting dark coloration to everything from fungi to frogs, and for many of the same reasons.

Melanin is a superb natural sunscreen. Because melanin molecules are composed of multiple units connected by strong carbon-carbon bonds, it has been difficult to precisely characterize the chemical compounds.[2] Nevertheless, scientists have been able to study the properties of naturally occurring melanins in some detail. Thanks to this research, we know that melanin is distinguished by a range of extraordinary optical and chemical properties. In the body, it is able to absorb, scatter, and reflect light of different wavelengths.[3] The melanins in human skin are a mixture of compounds, including melanin polymers, building blocks, and breakdown products. Together, these molecules can absorb all wavelengths of damaging UVR, protecting the vulnerable biological systems and molecular structures of the body, although melanin's ability to absorb solar radiation declines from the UV to the visible range.[4]

Melanin is made in melanocytes, the specialized immigrant cells described in chapter 1 (refer back to figures 1 and 2). These cells reside deep in the epidermis and in the matrix of the hair bulb. The story of melanocytes is a fascinating one with respect to both their phylogeny (their overall evolutionary history) and their ontogeny (their developmental history within each individual). Melanocytes originate in a part of the embryo called the neural crest, which flanks the neural tube (refer back to figure 18). They start out as actively dividing cells known as melanoblasts, which migrate to the epidermis during the eighteenth week of embryonic development, finding their way into the skin, ears, eyes, and brain covering (figure 20).

Melanocytes produce melanin pigment in small membrane-bound packets called melanosomes, which are then pushed out of the melanocytes and into the keratinocytes of the epidermis via spidery extensions known as dendrites. The size and shape of melanosomes, as well as the way they aggregate, influence their ability to protect the skin and underlying tissues from UVR. In darkly pigmented skin, the melanosomes are larger, melanin-rich, and dispersed evenly within keratinocytes. This arrangement permits them to absorb more energy than the smaller, less dense, and more lightly melanized melanosomes of lightly pigmented skin.[5] Additional protection

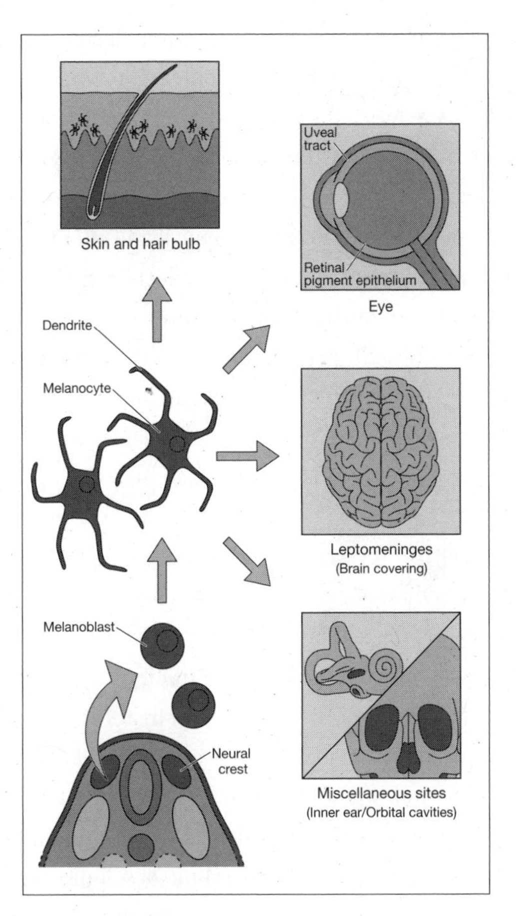

FIGURE 20. Melanocytes originate near the spinal cord in the developing embryo, at the neural crest, where they are known as melanoblasts. Early in the embryo's development, they migrate throughout the body to find their way into the skin, ears, brain, and eyes, where they produce pigment. (Illustration by Jennifer Kane.)

is offered by melanin dust, tiny particles of melanin in the epidermis not confined within melanosomes that are also capable of absorbing and scattering UVR.[6]

Recently, a new insight into the importance of melanosomes in human pigmentation has come from an unlikely source: a study of pigmentation in zebrafish. These small fish, originally from Africa, are common in both home aquariums and scientific laboratories. There are several varieties of zebrafish with different patterns of pigmentation, including a golden variety that has less melanin pigmentation than the wild type. The melanosomes

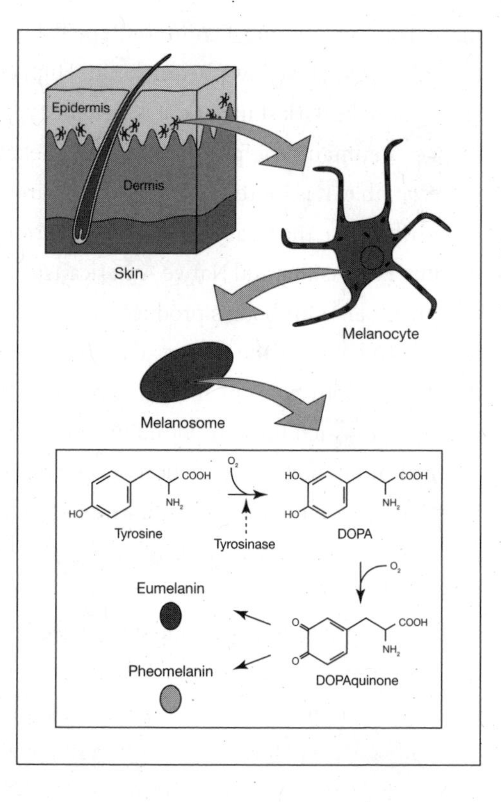

FIGURE 21. Melanocytes produce melanin, both the brownish-black eumelanin and the reddish-yellow pheomelanin, in small packets called melanosomes. Pigment-rich melanosomes are then transported into neighboring keratinocytes via the spidery dendritic arms of the melanocytes. (Illustration by Jennifer Kane.)

of the golden zebrafish are smaller and less dense than those of the normally pigmented type. A study of the golden zebrafish has shown that its characteristic melanosome structure and pigmentation are determined by a variant gene—and that the human equivalent of this variant gene is the predominant form in light-skinned people from Europe.[7] This similarity implies that the same genetic mutation created small melanosomes with reduced amounts of melanin pigment in Europeans. This variant gene and the light skin color it produced became widespread in what is often referred to as a "selective sweep." In other words, the light skin color the variant gene produces was so advantageous to survival when modern humans first occupied Europe that it quickly became the predominant type.

Two types of melanin are found in the skin of humans and all mammals. The first and most common type is brownish-black eumelanin; second is reddish-yellow pheomelanin. High concentrations of eumela are what make dark skin dark; it is also the type of melanin produced when you get a suntan. The presence of pheomelanin in human skin is much more variable. It is more common in red-haired northern Europeans, where it contributes to the total melanin composition of the skin. It is also found in some East Asians and Native Americans, but the amount appears to vary by individual.[8] Our bodies produce melanin in a chemical reaction involving the oxidation of the amino acid tyrosine using the enzyme tyrosinase (figure 21). Both types of melanin are manufactured through a common pathway, in which the compound dopaquinone is a key intermediary.[9]

The human body's production of melanin is governed by many factors, including pigmentation genes, hormones, and UVR. When the actions of genes and hormones are not in balance, an individual's melanin production can be completely or partly disrupted, causing the person to have little or no pigment in his or her skin, hair, or eyes—in other words, to become an albino.[10] The potential for albinism exists in all animals, including insects and other invertebrates, fish, birds, and mammals. Albino animals and people have a dramatically different appearance from those with normal levels of pigmentation (figure 22). Albinism is the normal condition for many species of fish and invertebrates that live in caves or at great depths in the oceans where solar radiation cannot penetrate. In these cases, there has been no strong pressure from natural selection to maintain production of protective melanin, and that ability has been lost, without ill effects.

Because the chemical pathways regulating melanin pigmentation are long and complex, and problems can occur at different steps, several types of albinism exist. In humans, two forms of this condition are recognized. Ocular albinism results from the loss of melanin production in the eyes only. Oculocutaneous albinism occurs when melanin production is prevented throughout the body, causing a lack of pigmentation in the hair, skin, and eyes. Individuals with oculocutaneous albinism who live in areas with high levels of UVR are exceedingly vulnerable to skin cancers. In South

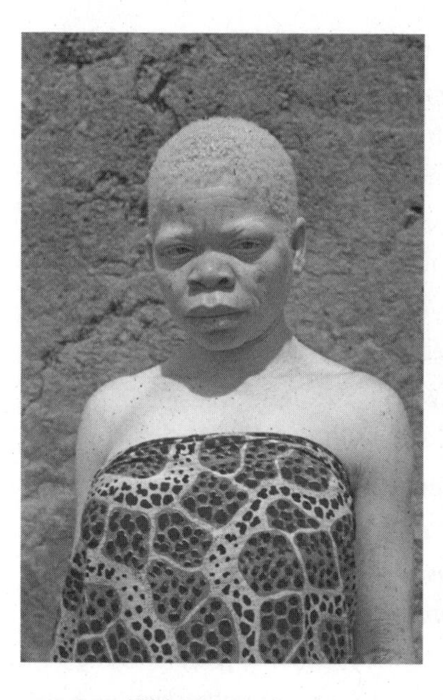

FIGURE 22. Albinism occurs when melanin production is interrupted or prevented because of a genetic or hormonal anomaly. The appearance of the albino gorilla Snowflake marks a dramatic departure from the dark colors characteristic of his species. In humans, albinism is more noticeable when affected individuals are born into a dark-skinned population, as was this woman of the Watembo tribe of the Congo. Albinos are less noticeable, however, when they belong to light-skinned populations; rock musician Johnny Winter, for example, is of probable northern European ancestry. (Photograph of Snowflake courtesy of the Barcelona Zoo. Photograph of Watembo woman courtesy of Edward S. Ross. Photograph of Johnny Winter © 2005 Robert Altman, www.altmanphoto.com.)

Africa, albinos of African origin suffer a thousandfold higher incidence of skin cancers than their darkly pigmented compatriots.[11]

In people with a normal range of pigmentation, the density of melanocytes varies over the surface of the body. Generally, the face and appendages are well-endowed, while the trunk is less so. Melanocytes are most highly concentrated in the groin, which accounts for the relatively dark color of the skin that covers the genitals, even in people with otherwise light skin. The total number of melanocytes is similar from one person's body to another, but not all melanocytes in all people are always actively producing melanin.[12] Some people with light skin produce very little melanin in melanocytes, while people with darker skin produce more. The manufacture of melanin can be stepped up in response to sun exposure, at least in people with lots of active melanocytes. This is the process that we usually refer to as tanning, and it is an important part of the body's protective response to UVR.

The number of active pigment-producing melanocytes also varies with age. Young children of both sexes have few active cells, but melanocytes begin to generate more melanin as puberty approaches. Females attain their darkest pigmentation at the onset of menstruation (varying from about age eleven through fourteen), whereas males continue to darken until their late teens. It is significant that when the skin color in any native population is examined, females are consistently lighter than males (although in some cases, this difference is not evident to the eye). As chapter 6 discusses in more detail, the greatest potential for melanin production in the skin in both sexes coincides with the human reproductive career. After the age of thirty-five or so, melanin production slows in both sexes. This is one of the reasons that older people tend to look lighter than their younger counterparts. Humans fade with age.

For many years, melanin in human skin was viewed only as a passive screening filter to protect us from UVR. We now know that melanin is not just a passive blotter—in fact, it actively participates in chemically neutralizing the harmful effects of UVR exposure. When melanin absorbs photons from solar radiation, it can undergo a chemical change itself. Recent evidence indicates that it is this chemical change that gives melanin its ability to "mop up" free radicals, potentially destructive compounds that are formed as intermediates in a wide variety of biochemical reactions in the body. Large quantities of free radicals can be created when cosmic radiation and UVR interact with the lipid molecules in cell membranes and other cellular components. Free radicals are extremely reactive chemically and are highly toxic to cells because they can damage DNA. Among the free radicals of greatest concern for biological systems are reactive oxygen species such as the superoxide anion and hydrogen peroxide.[13]

At the physiological level, then, melanin helps to protect us from the DNA damage caused by UVR and the free radicals it produces. It also helps to prevent UVR and other high-energy forms of radiation from breaking down essential vitamins such as folate.[14] As the following chapter explains, this latter function was of key importance in the evolution of human skin pigmentation.

When you look at the skin on your own body, you'll probably find that some areas, such as your face or the back of your hand, appear darker than others. Other places—for example, the inside of your upper arm—are quite a bit lighter. These areas of skin differ in color because of the different amounts of melanin they contain. The skin on the upper part of your inner arm has been very little affected by the environment over the years and has the least melanin in it. The color of this skin area represents your genetically determined baseline skin color, known as your constitutive skin color.

The parts of your body that are more routinely exposed to sunlight, such as your hands and face, are darker, having developed some degree of tan as a result of increased melanin production. This temporary darkening of the skin caused by exposure to the sun is referred to as your facultative skin color. Facultative pigmentation develops when UVR stimulates skin containing active melanocytes to produce melanin. This type of pigmentation is temporary: unless UVR continues to stimulate the melanocytes, the deeper color imparted by the extra melanin will be lost when skin cells are sloughed off. This is the mechanism that causes a tan to fade.

If you have had a lot of unprotected sun exposure, the differential between your constitutive and facultative skin colors will be great. This is true even if you are dark-skinned to begin with, since most skin types, dark or light, produce melanin in reaction to UVR. People who have naturally lighter skin, and who therefore produce less melanin, are, however, more likely to get sunburned, are less likely to tan, and are generally more susceptible to skin cancers.[15]

For years, scientists have worked to develop objective and reproducible methods for measuring skin color in living people. In the seventeenth and eighteenth centuries, verbal descriptions of skin colors—"white," "yellow," "black," "brown," and "red"—had to suffice. Such definitions have obvious problems. The actual colors that people associate with names of colors differ; what one person calls "light brown" may be another's "yellow." In the early twentieth century, these terms were replaced by less ambiguous methods of color matching, which involved the use of tablets or tiles of graduated color that were matched to the skin. The most popular such method was the von Luschan scale of skin color,[16] which was widely used by anthropologists through the 1950s.

Color matching methods were better than verbal descriptions, but they were still not satisfactory because they could not be consistently reproduced. There remained an element of subjectivity in assigning proper matches, and observers didn't always agree with one another. By the 1950s, interest in the study of human skin color was growing, and the need for a more objective way to measure and classify it became increasingly important. To meet this need, a method called reflectance spectrophotometry was introduced into anthropological fieldwork. This method had first been used to measure skin color in the late 1930s, but it became widespread in the 1950s with the development of a portable device that could be used in the field.[17]

The principle of reflectance spectrophotometry is easy to grasp. Light of different colors (and, thus, of known wavelengths) is focused on a small area of skin. The light reflected from the skin is then measured by a photocell. The reading from the photocell represents the percentage of light relative to a standard pure white block. Lightly pigmented skin reflects more light than darkly pigmented skin, and different wavelengths are reflected to different extents. Since the 1950s, anthropologists and dermatologists have devised numerous instruments for measuring skin reflectance, but the principle behind the operation of these devices is the same. Measuring skin reflectance remains the method of choice for the study of skin pigmentation because the procedure is standardized and lacks the subjectivity inherent in visual matching of colors.[18]

In medicine, the classification of skin color has stemmed primarily from the need to quickly and reliably evaluate the risk of skin cancer in light-skinned patients, in the setting of a doctor's office. Because lightly pigmented people differ in their ability to tan and are not equally susceptible to sunburn and skin cancer risk, the method of skin phototyping, developed in 1975, helps physicians accurately predict a person's reaction to moderate sun exposure. According to this classification system, there are six skin phototypes: three are referred to as "melanocompromised" (phototypes I–III) and three are considered "melanocompetent" (phototypes IV–VI). The definition of sun exposure in this system is thirty minutes of unprotected exposure without sunscreen at peak (summer) UVR levels.[19]

SKIN PHOTOTYPES	REACTION TO SUN EXPOSURE	SKIN COLOR
I	Burn and no tan	Pale white
II	Burn and minimal tan	Pale white
III	Burn and then tan well	White
IV	Tan no burn	Light brown
V	Tan no burn	Brown
VI	Tan no burn	Dark brown

Although skin phototyping has limited applicability, it has been widely embraced because it allows doctors to assess skin cancer risk in the office, without the need for fancy instrumentation. Individuals belonging to phototypes I, II, or III are more susceptible to skin cancers and sun-induced skin damage than those belonging to phototypes IV, V, or VI.

With this knowledge of the roles of UVR and melanin, we can begin to appreciate what all the varied colors of human skin actually mean to us as biological beings. The amounts of melanin in our skin were not determined randomly by some natural lottery, but by evolution through natural selection. We now need to see, step by step, just how this important chapter of human history unfolded.

6
color

As we look across the globe, humans display an astonishing variety of skin colors. In no other species do we find such a broad range. The reasons for this diversity are rooted in human evolutionary history and offer some of the most compelling examples of natural selection at work in the human lineage.

We can understand skin color in modern humans only by looking back at our evolutionary past. Earlier chapters have presented some of the key information, but it is worth reviewing a few points. Most species of our close primate relatives, the Old World monkeys and apes, have lightly pigmented skin covered with dark hair, which is probably the ancestral, or primitive, condition for the entire group to which these animals belong.[1] Their skin contains differing combinations of apocrine and eccrine sweat glands; our closest cousins, the African apes, have mostly eccrine glands, as humans do. From this information, we can deduce the ancestral condition of the skin of the last common ancestor of chimpanzees and humans, who lived around six million years ago: it was probably lightly pigmented, covered with dark hair, and endowed with a preponderance of eccrine sweat glands.[2] Significantly, this skin likely had the ability to develop more melanin pigment in response to sun exposure (in other words, to develop a tan) on those surfaces that were not covered with hair, such as the face and hands.

What happened to hominid skin after our lineage split from that of chimpanzees? During the period from six to about two million years ago, human evolution was confined to Africa and was dominated by species belonging to the genus *Australopithecus*.[3] The australopithecines—as members of this group are known—were made up of at least eight distinct species.[4] One of these species may have been the ancestor of the *Homo* lineage (or at least closely related to the direct ancestor), although this issue is still debated.[5] Before the evolution of the genus *Homo*, hominids as a whole exhibited shorter stature, more ape-like limb proportions, more ape-sized brains, and a generally more ape-like lifestyle. The australopithecines do not betray any anatomical signs that they engaged in the high-energy, long-distance sojourns that seem to have characterized *Homo*. Rather, they were walkers and respectable tree climbers, an ability that would have served them well during foraging or when they were threatened by predators.

The australopithecines were not yet naked apes. Their skin was probably very similar to that of the last common ancestor of chimps and humans—that is, light in color with a coat of dark hair. The skin of their faces and hands likely had the potential to develop darker pigmentation as the animals got older and were exposed to more sunlight, but in terms of their skin, they still looked like apes (see color plate 6).

The earliest known members of the genus *Homo* are recognized in the African fossil record beginning about two million years ago.[6] Compared to their australopithecine predecessors, these hominids had generally larger bodies, larger brains, and longer lower limbs. Their skeletons lacked the grasping toes that characterized their tree-climbing forebears and instead bore the long legs of powerful long-distance walkers and runners. The higher activity levels and longer daylight exposure we have been able to reconstruct for early *Homo* required that the skin of the species be essentially hairless and endowed with a high density of eccrine sweat glands in order to facilitate heat loss, as chapter 3 explains. This situation created a new physiological challenge for humans: protecting naked skin from the high UVR levels in equatorial sunshine. Dense hairy coats protect the skin of most mammals from UVR-induced damage because the hairs themselves absorb or reflect most short-wavelength solar radiation. Once most or all of the hair is lost, however, the skin is highly vulnerable. Because hominids at this stage lacked the technological wherewithal to protect themselves from the sun with clothing or shelter, they had to adapt biologically.

This adaptation took the form of a marked increase in the melanin content of the skin, and it represented a major change in the skin's appearance and function. Early members of the genus *Homo*—the ancestral stock from which all later humans evolved—were darkly pigmented (see color plate 7). This contention has recently been supported by genetic evidence demonstrating that strong levels of natural selection acted to produce dark pigmentation in these human ancestors.[7]

Why wasn't dark skin a liability in the heat? After all, dark objects absorb more radiation and become hotter in the sun than light ones do. Physiologists and anthropologists have carefully studied this topic over the years, mostly in investigations of human endurance under extreme environmental conditions. These studies demonstrate that darkly pigmented skin itself does not perceptibly increase the body's heat load in the face of intense solar radiation. Infrared radiation is the type primarily responsible for the heat buildup caused by sunshine, and dark skin and light skin absorb infrared radiation from the sun to nearly equal extents.[8] By far the most significant factors in increasing a person's heat load are external temperature, humidity, and the amount of heat the person generates as a result of exercise. So dark skin did not work against *Homo*'s ability to tolerate heat.

By the time the genus *Homo* emerged, then, about two million years ago, our ancestors' skin had become nearly modern in its color and hairlessness. Most of the evolution of the genus *Homo* was played out in Africa. Fossil evidence from Ethiopia shows clearly that the ancestors of modern

Homo sapiens existed around 155,000 years ago and that fully modern people evolved soon thereafter. Completely anatomically modern people started to make their way out of tropical Africa around 115,000 years ago, based on our knowledge of fossils from the Levant.[9]

By the time modern humans started leaving tropical Africa, they were culturally competent people. They made and employed a wide variety of tools; they used fire and cooked; and they had a sophisticated command of language. The pace of this modern human diaspora began to quicken around 100,000 years ago, but because so many of these migrations involved coastal routes (and now-inundated sites), our physical record of them is sparse; much of what we know comes from molecular evidence combined with information from bones and stones. These bodies of evidence show that fully anatomically modern people were established in southern Africa around 70,000 years ago, in Australia around 60,000 years ago, and in Europe around 50,000 years ago.[10] As modern people moved into these far-flung and environmentally diverse places, their bodies and cultures adapted to new circumstances. One of the most important adaptations had to do with their skin color.

What my colleagues and I have learned in our own research on the evolution of skin color is that melanin levels in the skin represent a compromise, which has been struck by evolution through natural selection.[11] This conclusion in itself is not new; what is new is the establishment of causal relationships between levels of pigmentation and human reproductive success. Reproductive success—survival to reproductive age, successful reproduction, and the survival of offspring—is the basis of natural selection and the final arbiter of evolution. In the past, most explanations for the evolution of dark skin focused on the protection it afforded from UVR effects that were detrimental to health but that did not necessarily affect reproduction—for example, sunburn, sun-related skin degeneration, and skin cancer. But these effects cannot be invoked as the primary drivers in the evolution of dark skin because they have little impact on an individual's reproductive ability.[12] Because reproductive success is what evolution is all about, adaptive explanations for a given trait must demonstrate some reproductive benefit.

UVR influences several chemical compounds in the body that are vital for successful reproduction, including DNA, folate, and vitamin D. Therefore, skin pigmentation should be dark enough to prevent or slow the breakdown of important biomolecules in the skin by UVR, but light enough to permit the production of other important biomolecules catalyzed by UVR. In other words, melanin is the governor.

This theory is based on two long-standing observations, introduced in chapter 4. The first is that longer-wavelength UVR (UVA) destroys the B vitamin folate. A lack of folate reduces reproductive success because it inhibits the production of DNA, which is needed for cell division. The second is that shorter-wavelength UVR (UVB) synthesizes vitamin D in the skin. A lack of vitamin D adversely affects reproductive success because it impairs the body's calcium metabolism. Different levels of skin pigmentation have therefore evolved to balance these somewhat contradictory needs.

Evolution has produced two opposing gradients, or clines, of skin pigmentation. The first cline grades from darkly pigmented skin at the equator to lightly pigmented skin near the poles, corresponding to the need for photoprotection. The second cline grades from lightly pigmented skin near the poles to darker skin near the equator, allowing UVR to penetrate the skin for vitamin D production. Between the extremes of these clines, we find people with moderate levels of genetically determined pigmentation who have enhanced ability to tan, depending on seasonal UVR levels.

If you live in a part of the world that receives very high levels of UVR, it is highly advantageous to have as much melanin in your skin as possible in order to protect your DNA and folate levels from the damaging effects of UVR. The extremely dark skin tones that we see among people in equatorial Africa have evolved to meet this need. Although dark skin provides excellent protection against the harmful effects of UVR, it also greatly slows the process of producing vitamin D in the skin.[13] Because melanin is such an effective sunscreen, people with very dark skin must spend more than five times as long in the sun as those with light skin in order to make the same amounts of vitamin D.[14]

As ancient members of our own species first moved out of equatorial latitudes, their exposure to UVR—and especially to UVB, which stimulates vitamin D production—was dramatically reduced. Under conditions of low UVR, their dark pigmentation tended to slow or prevent the process of synthesizing vitamin D in the skin. Therefore, as humans moved out of the tropics, natural selection promoted the lightening or depigmentation of their skin in order to facilitate the photosynthesis of vitamin D.[15]

Using remotely sensed data on UVR levels at the earth's surface and the precise dosage of UVB necessary to synthesize vitamin D in human skin at a specific latitude,[16] we can calculate and map the potential for producing vitamin D in the skin in various geographic locations. Near the equator, UVR levels throughout the year are adequate to manufacture vitamin D, although the rate is considerably slower in darkly pigmented people. As we move out of the tropics into the middle latitudes, from about 25° to 50°, there is not enough UVR to produce vitamin D in the skin of a lightly pigmented person during at least one month of the year. In latitudes above about 50°, in the far north, levels of UVR are much lower. When these levels are averaged over the entire year, they are insufficient to permit enough vitamin D in the skin of a lightly pigmented person to maintain normal health.[17] This was a real problem in human evolution. It is also a big problem today because the farther north you go, the harder it is for you to make vitamin D in your skin.

For people with darker skin who filter out a great deal of environmental UVB with their built-in melanin, these vitamin D zones have a different shape.[18] The "safe zone" for vitamin D production near the equator is narrower for people with dark skin than it is for people with light skin. At high latitudes, it is almost impossible for dark-skinned people to produce vitamin D during most of the year. A recent study of schoolchildren in South Africa that compared the vitamin D levels of darkly pigmented and albino children offered dramatic evidence. The darkly pigmented children had significantly lower levels of vitamin D and consequently required a substantially higher dietary intake of the nutrient to attain the same physiological levels as the albino children.[19]

Strong natural selection for adequate vitamin D synthesis in the skin was the main cause of the evolution of lightly pigmented skin in human populations at high latitudes. Modern humans had to become lighter as they moved into regions with lower levels of UVB. This same vitamin D imperative also likely affected earlier hominids living outside the tropics. The ancestry of the well-known Neandertals, who lived in Europe and western Asia about 300,000 to 30,000 years ago, can be traced to a European stock of early *Homo*. Neandertals are not closely related to modern Europeans,[20] but they inhabited many of the same places and experienced many of the same environmental conditions. It is therefore reasonable to infer that members of their lineage developed lighter skin in the course of their adaptation to life in low-UVR parts of Eurasia (see color plate 8), possibly through different genetic mechanisms than those that brought about skin lightening in the ancestors of modern Europeans. Although we cannot be certain, it is likely that Neandertals also became somewhat hairier in the course of their evolution from tropical antecedents. Thicker body hair would have provided a small measure of extra warmth to their naked skin. Neandertals survived the inimical climatic conditions of the last ice age largely because of their cultural capabilities: they took advantage of natural shelters and used fires and simple body coverings made from animal hides to stay warm.

Like their modern human counterparts in similar latitudes, Neandertals probably had an ability to tan when UVR levels were high. In regions where UVB levels fluctuate dramatically with the seasons, evolution of the ability to become temporarily darker during periods of higher UVR intensity and then fade as UVR levels fell would have been favored. This explains why the native peoples of the circum-Mediterranean area and others inhabiting the zone between approximately 23° and 40° latitude have excellent

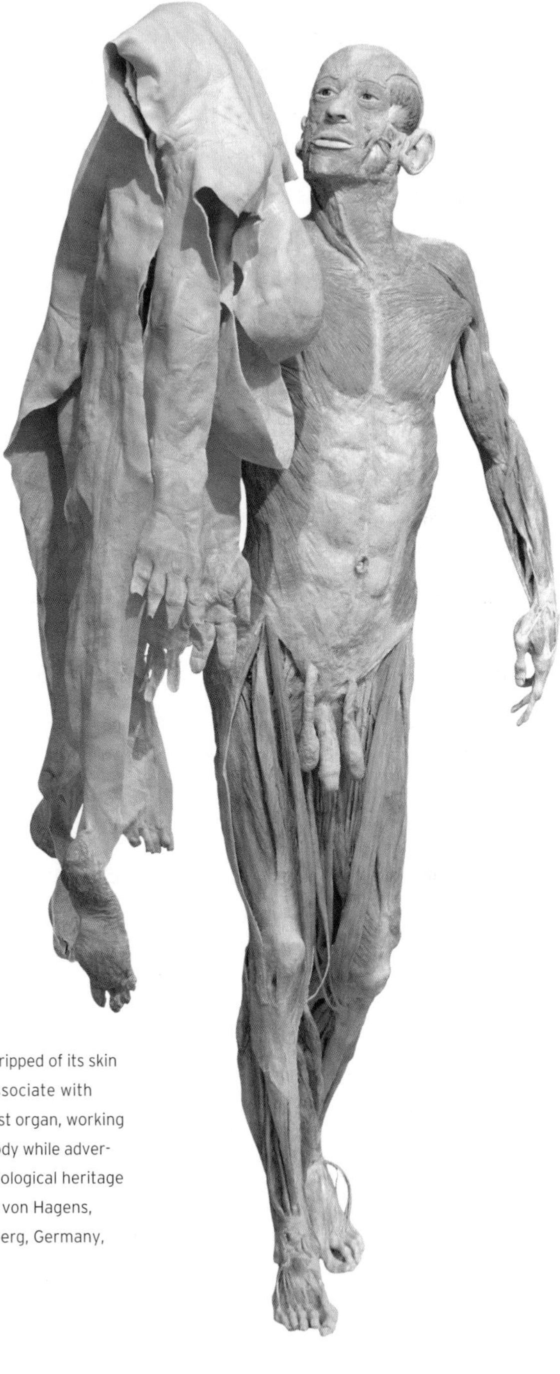

PLATE 1. An *écorché.* The body stripped of its skin lacks most of the qualities we associate with personhood. The skin is our largest organ, working to preserve the integrity of our body while advertising important aspects of our biological heritage and cultural identity. (© Gunther von Hagens, Institute for Plastination, Heidelberg, Germany, www.bodyworlds.com.)

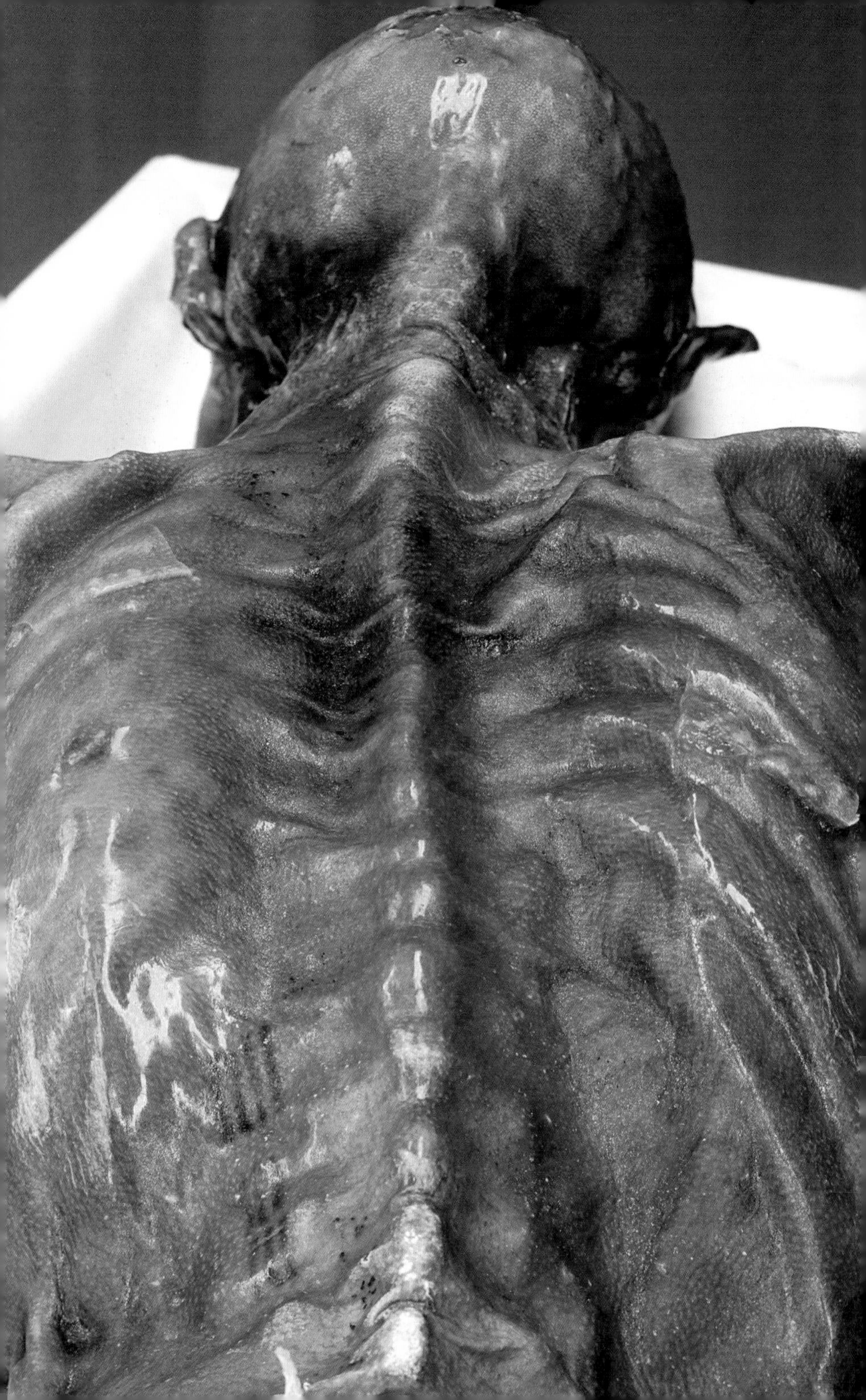

PLATE 2. The mummified skin of Ötzi, the Neolithic Iceman retrieved from an alpine glacier, is in good condition thanks to nearly 5,000 years of preservation in a natural icebox. It is the oldest preserved human skin known. Continued refrigeration prevents Ötzi from decomposing further. Considerable speculation has arisen about the curious tattoos that appear on his skin, a few of which are visible on the left side of his back. (© Marco Samadelli, South Tyrol Museum of Archaeology.)

PLATE 3. The facial skin of Tollund Man was remarkably well preserved in the cold, acidic environment of a peat bog, which inhibits decomposition. Tollund Man, who lived during the fourth century BCE, was recovered from a peat bog on the Jutland Peninsula of Denmark. (© Silkeborg Museum, Denmark.)

PLATE 4. The naked mole rat spends its life burrowing through warm ground in subterranean colonies in the African tropics. Its nearly hairless skin has evolved as a way to help the animal maintain an even body temperature in its underground environmen (© Jesse Cohen, Smithsonian's National Zoo

PLATE 5. The hippopotamus is a large mammal that stays in the water during the heat of the day and moves onto land at night to feed. It loses water through its skin in order to dissipate heat, in a process called transepidermal water loss. In this photo, beads of the hippo's unique "red sweat," which acts as both a sunscreen and a moisturizer, can be seen on its brow. (Courtesy of Kimiko Hashimoto.)

PLATE 6. A reconstruction of the hominid species *Australopithecus afarensis*, by Mauricio Antón. This species lived in eastern and northeastern Africa about 3.5 million years ago and probably had light skin covered by dark hair, similar to that of the common ancestor of chimpanzees and humans. (© 2005 Mauricio Antón.)

PLATE 7. A reconstruction of the hominid species *Homo ergaster*, by Mauricio Antón. This species is known from sites in eastern Africa dating from about 1.8 to 1.6 million years ago. These hominids probably had mostly hairless bodies covered with dark skin, which scientists believe was the ancestral condition for the genus *Homo*. More active than their predecessors, especially during hot daylight hours, they evolved dark skin to protect them against UVR and lost their hair to facilitate heat loss. (© 2005 Mauricio Antón.)

PLATE 8. A reconstruction of a Neandertal man from central Europe, by Mauricio Antón. The ancestors of Neandertals, who emerged about 300,000 years ago, and other ancient hominids of Europe probably evolved lighter skins than their African forebears because of the lower UVR levels there. They may have also subsequently regained some of their body hair, as suggested in this reconstruction. (© 2005 Mauricio Antón.)

PLATE 9. This San (Basarwa) man from Botswana exhibits the moderately pigmented skin characteristic of the indigenous peoples of southern Africa. Like the peoples of Eurasia, southern Africans have undergone depigmentation in the course of their evolution. (Courtesy of Edward S. Ross.)

PLATE 10. The dark skin of this Pokot woman from Marigat in central Kenya is typical of many dwellers of the arid African tropics. (Courtesy of Edward S. Ross.)

LATE 11. The Berbers of Tunisia have
ıoderately pigmented skin that tans well but
uffers damage from long-term sun exposure.
he deep wrinkles on the face of this Berber
voman from Chenini, Tunisia, result from a
reakdown of connective tissue in her skin,
rought about by too much UVR exposure. (©
Vinston Yeung/www.yeungstuff.com.)

PLATE 12. The Sámi people of Finland exhibit
extremely lightly pigmented skin, an
adaptation to the low UVR levels that prevail
near the Arctic Circle. (© Fred Ivar Utsi
Klemetsen.)

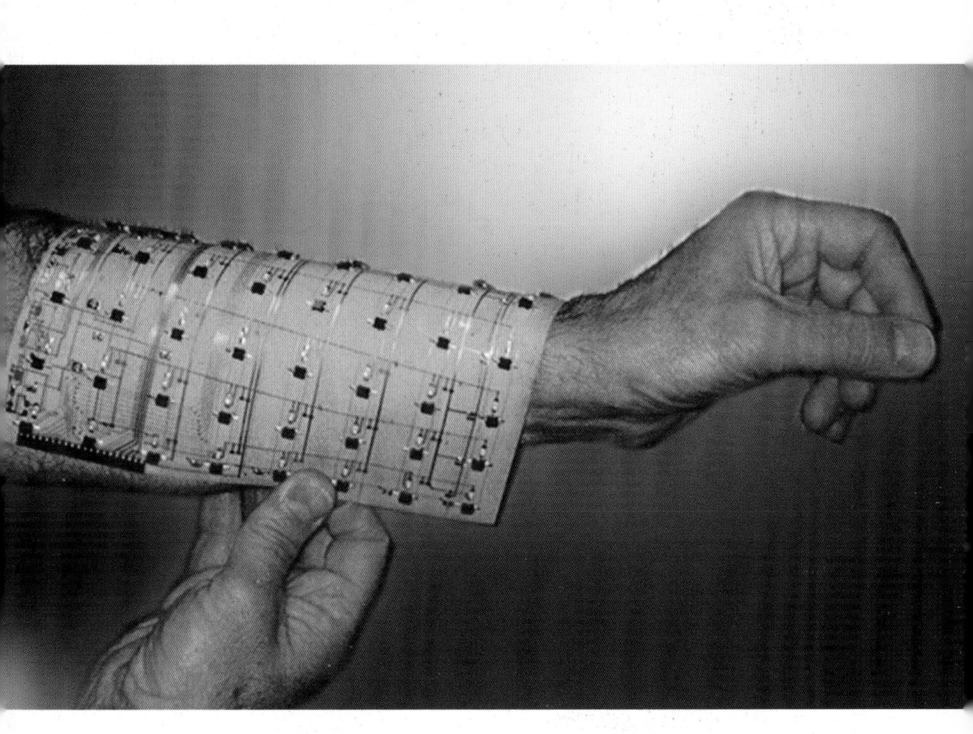

PLATE 13. In remote parts of northeastern Myanmar, facial tattooing of women was common until very recently. Among the Dulong, young women were tattooed at the age of twelve or thirteen with a design that was distinctive to their region. Some groups use this marking to deter outsiders from luring away desirable women. (© 2005 Dong Lin.)

PLATE 14. Artificial electronic skin is in an early stage of development. The hope is that it will eventually be able to provide robots of the future with a sense of touch and, perhaps, a sense of self. (© 2005 Vladimir Lumelsky. *Sensing, Intelligence, Motion: How Robots and Humans Move in an Unstructured World.* John Wiley and Sons.)

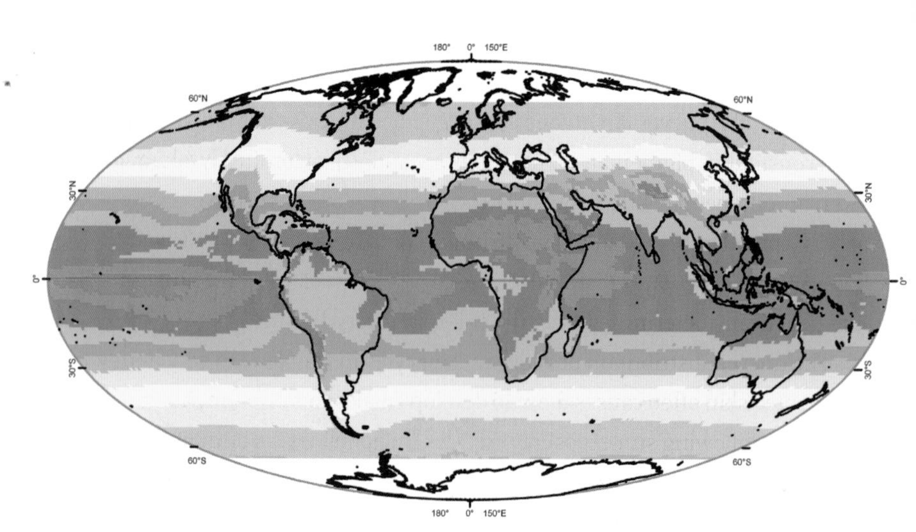

MAP 1. Annual average UVR levels received on the earth's surface, based on data from the NASA TOMS satellites (version 7). Levels of UVR are highest within the tropics and especially at the equator (red and blue areas), although latitude is not the only determinant. Extremely high northern and southern latitudes (gray areas) receive very low levels, and then only at the peak of summer. (© 2005 George Chaplin.)

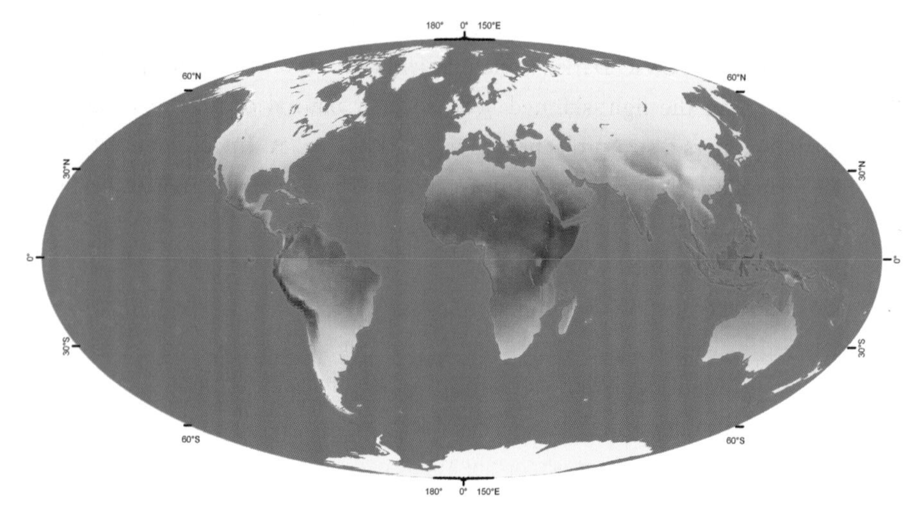

MAP 2. Predicted skin color for modern humans, based on the relationship between measured skin reflectance data and environmental factors, primarily UVR levels. The variations in color approximate actual skin color. The darker skin tones found near the equator and in the tropics grade to paler tones at higher latitudes. (© 2005 George Chaplin.)

tanning ability. People with very light constitutive pigmentation (skin phototypes I and II) never tan or tan very little, whereas those with moderate to dark constitutive pigmentation (phototypes V and VI) tan profusely (although even among people with ostensibly similar levels of constitutive pigmentation, considerable variation in tanning potential exists).[21] In all people, facultative pigmentation fades over the course of weeks and months until the genetically determined baseline color is restored.

The uneven distribution of active melanocytes in people with very lightly pigmented skin often causes them to develop small spots of pigmentation—freckles—following sun exposure. Freckles (or ephelides, as dermatologists call them) occur more frequently in very fair-skinned individuals; they tend to appear early in childhood and mostly disappear with age. Freckles darken and fade with seasonal changes in UVR exposure much as a tan does. Because people with freckles are at much greater risk of developing skin cancer than those without them, they need to reduce their UVR exposure as much as possible by routinely using sunscreen and covering up when outside.[22]

Relatively light-skinned people are not the only ones who tan; tanning is also an important aspect of how darkly pigmented people have adapted to high levels of UVR. Dark-skinned individuals can tolerate longer sun exposure than the light-skinned because their natural complement of melanin confers a sun protection factor (SPF) of 10–15. In contrast, the SPF of moderately pigmented skin, such as we find in southern Europe or central Asia, is only 2.5.[23] For most people, including those who are naturally dark-skinned, exposure of the skin to peak UVR will make their skin darker because UVR stimulates their melanocytes to produce more melanin.[24]

The process of developing a tan involves immediate tanning followed by delayed tanning, which can take more than forty-eight hours to become visible. What may be most important about naturally dark skin is that its heavily pigmented melanocytes have a greater capacity to resume normal cell division after UVB exposure than their lightly pigmented counterparts, perhaps because the cells suffered less damage to their DNA.[25] Although

vier pigmentation protects DNA, UVB nonetheless adversely affects dark
ı because it disturbs the skin's immune system. UVB damages the Langerhans cells of the skin, regardless of the level of pigmentation, and thus weakens the skin's ability to protect itself from potentially harmful microorganisms and substances in the environment.[26]

Many people think that tanning protects them from the harmful effects of further UVR exposure, but this is a dangerous fallacy. For people with genetically light skin, tanning does not significantly increase the SPF sufficiently to protect DNA from UVR-induced damage. Repeated exposure of tanned skin to UVR increases the number of active melanocytes and the intensity of melanin production, but the increased melanin concentration in the tanned skin of a normally light-skinned person does not approach the photoprotection conferred by natural melanin in intrinsically darker-skinned people. People with lightly to moderately pigmented skin who routinely expose themselves to UVR produce free radicals in their skin that cause structural proteins in the dermis to break down, which leads to premature aging of the skin and, over time, to visible wrinkles and uneven pigmentation.[27]

Armed with a basic understanding of how human skin pigmentation evolved, we can take an imaginary "tour" to see how skin color varied in the early millennia of our species. It's particularly interesting to roll back the clock to a time about ten thousand years ago, before the so-called agricultural revolution, when people started moving long distances more quickly. Taking a "walk" along the great length of the meridian 20° east from Greenwich allows us to glimpse almost the entire range of skin color variation that exists among humans. Along this narrow swath, which begins at the Cape of Good Hope at the southern tip of Africa, we first encounter the native gathering and hunting peoples of the Kalahari Desert, who have moderate amounts of melanin pigment in their skin (see color plate 9). Heading north through the Kalahari, we pass over the Tropic of Capricorn and into the territories of more darkly pigmented gathering tribes, who are related to the people of the far south. Pressing toward the equator through the Congo Basin, we encounter people with heavily pigmented skin. We encounter numerous groups of very dark-skinned people as we move northward through the Sahel and the Sahara Desert toward the Tropic of Cancer (see color plate 10).

As we enter the Libyan Desert, people are perceptibly lighter in color and become even more so as we approach the southern shore of the Mediterranean (see color plate 11). Like the dwellers of the southern Kalahari, the peoples of the circum-Mediterranean are moderately pigmented and are able to develop deep tans after lengthy exposure to sunlight. After crossing the Mediterranean, we enter the Balkan Peninsula and proceed north over the Plain of Hungary into the Carpathians and then onto the North European Plain. Here, people are significantly lighter, although they can still develop noticeable tans in the summer months. Entering the Baltic Sea, we ply the Gulf of Bothnia before landing on the southern shore of Scandinavia, with its pale-skinned inhabitants who are much lighter in color than those on the southern shore of the Baltic. As we approach the Arctic Circle and Lapland, we enter the sparsely inhabited homeland of the Sámi people, who have very lightly pigmented skin (see color plate 12).

What is clear from our journey over this elongated trail is that, from one end to the other, skin color changed mostly very gradually from one latitude to the next. There was no noticeable break or sharp discontinuity along this road of skin, just innumerable shades of brown, tan, and off-white. Today, the natural gradient of skin color from equator to poles has been disrupted somewhat as humans have migrated to distant lands at faster paces, especially in recent centuries, but it is still one of the most obvious patterns of human biological variation.

As humans from established cultures began to take long journeys overland and by sea for purposes of mapping and trading in relatively modern centuries, they were sometimes startled to discover different skin colors. Even in travelers' accounts from fifteenth-century Europe, the skin colors of newly encountered peoples are reported as surprising.[28] As natural historians and geographers—mostly from Europe—ventured into Asia, Africa,

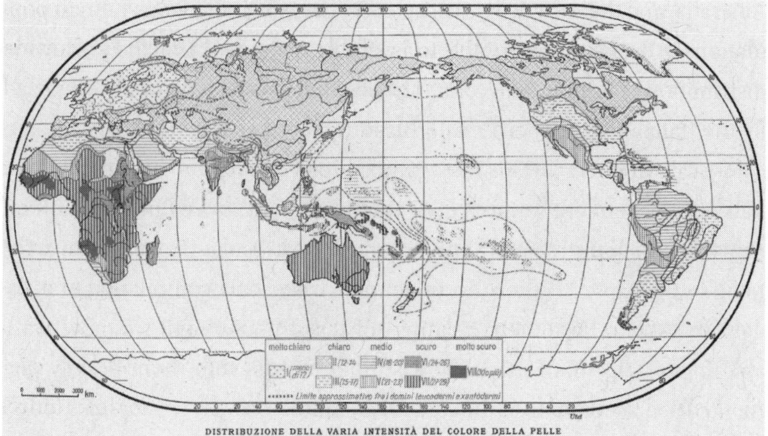

FIGURE 23. The most commonly reproduced map showing human skin colors was composed by the Italian geographer Renato Biasutti and published in 1959. This map is based on data compiled from a variety of sources and includes some extrapolations for areas for which no information on skin color was available.

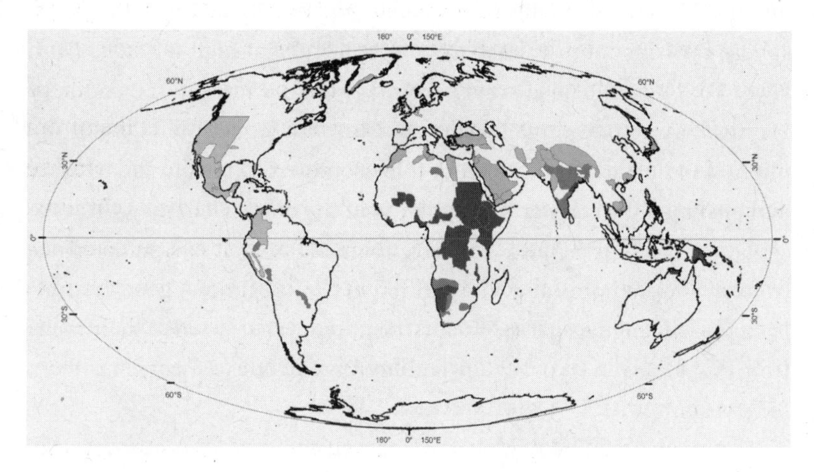

FIGURE 24. This map, based on actual skin reflectance data for indigenous populations, is more accurate than Biasutti's (shown in figure 23). Both maps depict similar trends, however, showing a preponderance of darker-skinned people near the equator, where UVR levels are highest, and lighter-skinned people in higher latitudes, closer to the poles. (© 2005 George Chaplin.)

Australia, and the Americas and began to study the indigenous human populations in detail, they were able to assemble maps depicting the worldwide distribution of human skin color. The best-known such map was composed by the Italian geographer Renato Biasutti, based on the von Luschan skin color scale (figure 23). This map gained broad circulation during the latter half of the twentieth century, even though Biasutti lacked information on skin color for some regions and was forced to extrapolate from known areas to fill in the gaps.[29] A more accurate map, based on a compilation of standardized skin reflectometry readings of indigenous peoples, is now available (figure 24). Both maps show similar trends, with more heavily pigmented people found near the equator and incrementally lighter ones found closer to the poles.[30]

One of the most interesting aspects of the map shown in figure 24 is that the percentage of dark-skinned people in the Southern Hemisphere is greater than in the Northern Hemisphere. There is in fact a bias in the distribution of land masses relative to the equator.[31] The Northern Hemisphere contains more habitable continental area at high latitudes (with lower UVR doses), and a larger percentage of land mass in the Southern Hemisphere is concentrated at the equator (with higher UVR doses). You might also notice that there seem to be some exceptions to the trend of darker skin at the equator and lighter near the poles. Generally, these exceptions represent peoples who have migrated to their current locations from an ancient homeland with a different UVR regime. A good example is the Bantu-language speakers of Africa, who have expanded their range from equatorial west Africa into southern Africa over the past four thousand years.

The regular gradation of skin color by latitude exists because levels of UVR at the earth's surface are strongly correlated with latitude and these, in turn, are correlated with skin pigmentation. Detailed study of skin pigmentation and UVR levels has shown that skin color is most closely related to UVR levels in the autumn, when UVR levels are near their lowest of the year. This correlation may exist because skin color is most strongly constrained by low rather than high UVR levels and the dampening effect that low UVR has on vitamin D production.[32]

Using the relationship that exists between skin color (as measured by skin reflectance) and environmental parameters including UVR, it is possible to construct a map of predicted human skin colors, in which the colors are fairly realistic approximations of the true color of skin (see color map 2). This map depicts an idealized situation in which we assume that humans worldwide have inhabited their respective regions for similar amounts of time and have followed similar cultural practices that could affect their skin color. It's important to remember when looking at such a map that the world's indigenous peoples haven't all been in their home countries for the same length of time. Also, human populations differ in how they cope with the sun. Thus, the actual skin colors of some peoples don't closely match the colors predicted for their current habitats.

As scientists have studied human skin color over the years, one consistent observation is that women have lighter skin color than men. This is true for all indigenous peoples, even for those who are very dark-skinned, among whom such differences are not readily visible.[33] Some have speculated that lighter skin evolved in females to mimic the paler skin of infants, who in all populations have the lightest skin. By imitating the infant condition, this argument goes, females could garner some measure of the same social protection afforded to infants in groups that include stronger and potentially hostile males. Others have maintained that lighter skin in females can be traced to a history of conscious choices by males who prefer more lightly pigmented females as mates, possibly because of the association between light skin and infancy.[34] These hypotheses were based on observations that the attraction between human infants and human females results in part from their lighter pigmentation, as well as from other features of their appearance. According to this reasoning, lighter-colored females are perceived as more feminine than darker females and thus are preferable as sexual partners. Male-female differences in skin color may also be explicable from a male perspective: natural selection may have favored darker skin in males to optimize folate levels in the body, which would safeguard sperm production, a process that depends on folate for DNA synthesis.

I have advanced a different explanation of this sexual dimorphism, or genetically determined difference between the sexes, in skin pigmentation—and it involves vitamin D. During a woman's reproductive years, she is geared up not only to sustain the calcium reserves needed to maintain her own body but also to build the reserves of her offspring. During pregnancy, and to even a greater extent during breastfeeding, a woman's need for calcium is nearly double that of a man of the same age. The calcium and phosphate stored in her bones are mobilized on a large scale to help build the skeleton of her fetus and newborn child,[35] which means that she urgently needs to replenish the calcium stores in her own bones through increased intake and absorption of dietary calcium. If vitamin D is in short supply, she cannot absorb calcium, and her bones and those of her offspring suffer accordingly. In serious cases of vitamin D deficiency, the bones of the newborn baby don't harden properly, leading to the tragic deformity of rickets. The mother, although less obviously affected, suffers thinning and softening of the bones (osteomalacia) as a result of mineral depletion, leaving her with a weakened skeleton and a heightened long-term risk of bone fractures.

To avoid these problems, evolution has acted to ensure that mothers make enough vitamin D in their skin by selecting for females who are slightly lighter in color than males. By having lighter skin, females can produce slightly more vitamin D than males under the same UVR conditions, optimize their absorption of calcium, and improve their chances and their infant's chances of healthy survival and reproduction. Females maintain a delicate balance with respect to natural selection. They must have sufficiently dark skin to protect their folate and DNA, but light enough skin to maximize vitamin D production. This is evolution at its finest, establishing an effective biological compromise to ensure survival of the species.

These physiological arguments do not preclude sexual selection as a factor in creating the patterns of sexual dimorphism in skin color that we observe in human populations today. It seems unlikely, however, that sexual lection alone accounts entirely for these patterns because females are lighter than males in all populations studied, even those in which the difference can be detected only by instrumentation, not by the naked eye. But it may well be that male preference for lighter females in many populations has accentuated the preexisting disparity in skin color that was initially established by natural selection.[36] Many societies are known to express preference for lighter-skinned women, a view promoted by the vigorous marketing of skin lightening creams in both industrialized and developing countries.

Men and women attain their darkest levels of skin pigmentation during their early reproductive years, for reasons that make considerable evolutionary sense, considering the importance of an effective melanin shield in protecting the body's folate and DNA. In addition, however, females develop darker pigmentation on some parts of their bodies early in pregnancy. This phenomenon, known as melasma or chloasma, is characterized by darkening of the nipples and areolae and—to lesser extents—the abdominal wall, genitalia, and face. The extra pigmentation on the cheeks, nose, and forehead that is the hallmark of melasma is sometimes referred to as the "mask of pregnancy." Melasma increases throughout pregnancy, as the melanocytes step up melanin production and as the number of melanocytes in the skin actually increases.[37] The darkening of a pregnant woman's nipples and areolae is permanent and intensifies with successive pregnancies.

Blotches of dark pigmentation on a woman's face in a pattern similar to that of melasma can be a side effect of long-term oral contraceptive use, apparently because facial melanocytes are specifically sensitive to the estrogen and progesterone contained in birth control pills. Less well established is the reason for melasma-like changes on the face and body in response to the hormonal changes of the menstrual cycle in some women.[38] Many women complain of dark circles under the eyes and around the mouth during their period, perhaps as a result of temporarily increased production of melanin in facial melanocytes. It is tempting to speculate that the facial darkening of melasma may be an evolutionary adaptation to protect females of reproductive age from UVR. The ability of the melanocytes of the face and breast to respond quickly to hormonal changes suggests that additional protection for these regions benefits female health and would have been favored by natural selection.

Clearly, the most important determinant of human skin colors has been adaptation to UVR. But in the course of human history, numerous other factors have come to be influential. Both our culture and our mobility, for example, have shaped the action of natural selection over time. During the days of human prehistory, people had fewer cultural trappings to protect themselves against the environment. Under these conditions, biological adaptations to the environment evolved, including changes in skin color, body proportions, and methods of regulating thermal cooling. Over time, however, the weight of "cultural capital" has increased,[39] strengthening our ability to protect ourselves from the vicissitudes of the environment. With greater cultural competence, cultural solutions to the environmental challenges of heat, cold, and UVR exposure—such as wearing clothes and constructing shelters—have largely come to supplant biological ones.

In addition, the skin color of a population today is related to how long it has inhabited a particular area and how far—in terms of latitude—it has moved from its ancestral homeland. There appears to be a lag period between the time a group settles in an area and the time that it reaches its optimum skin color for the UVR conditions of that area. We don't yet know the length of this lag time, but it is related to the intensity with which natural selection operates on the population. In this connection, it is interesting to examine the skin colors of the native populations of equatorial South America. These New World populations have long been recognized as lighter-skinned than their counterparts at similar latitudes and altitudes in the Old World. Their lighter skin may be attributable to a combination of how recently they migrated to South America from Asia (within the last ten to fifteen thousand years) and the cultural practices and accoutrements they possessed, such as different types of clothing and shelter, that protected them from high levels of UVR exposure.

Another factor that has come into play, especially in the past two hundred years of rapid and wide-ranging human travel and migration—at a pace previously unknown in human history—is interbreeding between once widely separated and disparate peoples, resulting in offspring of intermediate skin color. This development is particularly evident in large urban centers on most continents and in countries such as the United States that have promoted immigration for many generations.

Diet has also played a part in the story of human skin color, with the role of dietary vitamin D assuming particular importance. Take the case of the Eskimo-Aleut peoples of the northeast Asian and North American Arctic. They exhibit darker skin tones than we would predict on the basis of the low UVR levels in their native habitats. Significantly, however, the UVR received in these regions consists almost exclusively of UVA throughout the year, with virtually no UVB except for extremely small doses in the summer, which seriously reduces vitamin D synthesis.[40] Living at this latitude has been possible only because the Eskimo-Aleut people subsist largely on a diet that is extremely rich in vitamin D. In fact, the aboriginal Eskimo-Aleut diet consists primarily of vitamin-D-rich foods such as marine mammals, fish, and caribou.[41] It is interesting that, with natural selection's pressure toward lighter skin apparently relaxed because of their diet, Eskimo-Aleuts have evolved darker skin to protect themselves from the high levels of UVA that reflect on them from snow, ice, and open water.

Our discussion has so far focused on the probable evolutionary forces responsible for different skin colors in humans. These forces act on a person's constitution and behavior, which are parts of that individual's phenotype. Behind every phenotype, though, is a genotype—the genetic basis for the external appearance. The genetic basis of human skin coloration is not yet well understood, and science is only now beginning to explore it vigorously, as one of the many intellectual offshoots of the Human Genome Project and the field of comparative genomics. Studies of the genetic basis of skin color in humans have been built on many studies of the genes that regulate the pigmentation of coat color in other mammals, especially mice.[42] So far, 60 of the 127 currently recognized pigmentation genes in the mouse appear to have functional counterparts, or orthologs, in humans.

Human skin pigmentation is not a simple trait specified by one gene or set of genes. Rather, it is determined by the synchronized interaction of many genes with the environment. Because of this complexity, scientists have found it extremely challenging to identify the relative roles of variant genes and varying environments in producing different skin color phenotypes. So far, most attention has focused on a gene known as the melanocortin-1 receptor *(MC1R)*. This is one of the major genes involved in determining human hair and skin pigmentation, through its role in controlling whether the melanocytes will produce eumelanin or pheomelanin. (Chapter 5 described these two types of melanin; eumelanin darkens the skin.) The *MC1R* locus, or gene, shows little variation within Africa, suggesting that there is strong selective pressure (sometimes referred to as "purifying selection") to maintain the gene's function in enabling eumelanin production. Outside Africa, the gene is characterized by high levels of variation (polymorphism). Variant forms, or alleles, of the *MC1R* gene in northern European populations are associated with red hair and fair skin, a greatly reduced ability to tan, and a high risk of skin cancers. The global pattern of variation in the gene suggests that adaptive evolution for sun-resistant forms of the *MC1R* genes began when humans first became hairless in tropical Africa and that human movement into the less sunny climes of Eurasia favored any mutant *MC1R* allele that did not produce dark skin. This interpretation is controversial, and research into the functions of the various forms of the *MC1R* gene is still in its early stages. What is clear now is that the *MC1R* gene has been the object of purifying selection and has played an important role in maintaining dark pigmentation in Africans.[43]

Much remains to be learned about the levels, effects, and interactions of the variant genes influencing skin color. While the production of high levels of eumelanin appears to be under tight genetic control as a result of natural selection in regions of the world with high levels of UVR, we have increasing evidence that variation in the *MC1R* gene (and possibly others) is an adaptive response to natural selection for different forms of particular genes in different environments. Geneticists have long sought to find a gene or genes responsible for light pigmentation in human populations outside the tropics. We are now closer to that goal with the discovery that the gene governing a light-color variant of zebrafish has a counterpart in the genome of European people, as chapter 5 describes.[44]

From what we know about the movement of groups of early *Homo* species and of *Homo sapiens* in prehistory, and the timing and nature of such movement, it is clear that populations of humans have moved in and out of regions with different UVR regimes over the course of hundreds of thousands of years. It is possible that natural selection favored the evolution of either dark or light skin color in disparate places at different times, so that dark and light skin phenotypes could have evolved independently and may have changed within the same population, whose skin might have become repigmented or depigmented.[45] In other words, dark skin and light skin evolved more than once in the course of human development, as populations moved into regions of high and low UVR. This phenomenon would have been pronounced in the early history of the genus *Homo* (including the early history of *Homo sapiens*) when cultural buffers against the environment were less effective and sophisticated than they are today.

The evolution of different skin colors is one of the most engaging and important stories in all of human evolution. But the issue of human skin color involves much more than just a story about evolution. Because skin color is the most obvious way people vary, it has been the primary characteristic used to classify people into purportedly genetically distinct geographic groups, or "races." But when we consider the biological basis of skin color as we now understand it, this method of classifying people doesn't make sense. Skin pigmentation is clearly adaptive, and its evolution in specific populations has been strongly influenced by the environmental conditions of specific places—in particular, the levels of UVR. Dark skin and light skin evolved repeatedly in early phases of human evolution, as populations moved through diverse environments. Skin color, like the rest of our bodies, is a product of evolution by natural selection, based on the solar environment of our ancestors.

But such highly adaptive characteristics are of little use for classifying organisms into different species or races, because they are subject to parallelism, or convergent evolution—that is, similar appearances evolve because natural selection works to produce functionally comparable adaptations in comparable environments. The variant zebrafish gene with its counterpart among European peoples is an excellent case in point. This gene appears to have contributed to skin lightening only in Europeans. The depigmentation that has occurred elsewhere in the world—in northern Asia and southern Africa, for example—appears to have been driven by different genetic means. Dark skin or light skin, therefore, tells us about the nature of the past environments in which people lived, but skin color itself is useless as a marker of racial identity.

Differences in appearance, especially differences in skin color, have contributed to the development of ideas about "race" and "ethnicity" that have often included the belief that significant inherited differences distinguish humans.[46] This is despite the fact that our species exhibits less genetic differentiation among geographically distributed human populations than is observed in many other mammalian species. In recent years, an increasing number of studies have used skin color in medical research as a surrogate for "race" or a genetically distinct group. This approach is disturbing because it ignores the role of sociocultural factors in mediating the relationship between skin color and various disease processes. Skin color is not an accurate proxy of ancestry and must be used with great caution in medical circumstances when decisions are being made about patient treatment.[47]

Although skin color does not equal "race," it does have relevance for health. Many diseases afflict people because of their skin color phenotype, that is, because of the lightness or darkness of their skin, independent of other factors. For instance, light-skinned people regardless of their ancestry are more vulnerable to skin cancers when they are exposed to high lifetime doses of UVR, usually because they are living in new environments far from their ancestral homelands.[48] Similarly, rickets and other diseases (including colon, breast, prostate, and ovarian cancers) that are influenced by levels of vitamin D in the body are increasing among dark-skinned people of all ancestries who are chronically vitamin D-deficient because they live in UVR-poor regions far from their homelands.[49]

We can now fully appreciate the momentous role that skin color has played in our evolutionary history. Far from being a trivial aspect of our external appearance, skin pigmentation turns out to be a key mediator of our interactions with the environment, especially with the sun. As early members of our species moved around the world tens of thousands of years ago, our skin color responded by becoming darker or lighter as needed.

These days, things are quite different because we can move great distances very quickly, often to places far from the lands of our ancestors, new locations to which our skin is poorly adapted. Our skin doesn't have a chance to catch up, and we may find ourselves too light or too dark for our new environments. We rarely stop to think that the body we carry around with us is basically a Paleolithic model that evolved tens of thousands of years ago; most of the newfangled extras we've added in recent history have been cultural, not biological. Our skin color is part of our biological legacy, telling a precious story about the environment in which our ancestors lived. For this reason, we need to take our color seriously, determine how well matched it is to the place where we live, and adjust our behavior accordingly. Most of all, we need to be grateful for the many ways in which our particular skin color works to keep us healthy.

7
touch

The skin houses the body's most ancient sense, touch. Sometimes referred to as the "mother of the senses," touch has not garnered the scientific or public attention it deserves, possibly because its influences on human well-being are more subtle than those of the so-called distant senses of sight and hearing. Likewise, the role of skin as the medium of touch in primate and human evolution has gone largely unrecognized. For those endowed with all five senses, touch certainly is underrated.[1] And yet touch is central to primate experience, and has been for tens of millions of years. It has influenced our evolutionary history, and it profoundly affects us during the course of our lives. The importance of touch in human affairs is reflected in our language; notice how we use the word in common figures of speech dealing with communication: we ask our friends and family to "keep in touch"; and when someone or some thing stirs our emotions, we exclaim, "How touching!" or humbly state, "I am touched."

Touch involves stimulation of the skin by mechanical, thermal, chemical, or electrical means and the resulting sensations of pressure, vibration, temperature, or pain. Primates are the most touch-oriented of all mammals. Tactility in primates evolved at the dawn of the group's history, when ancient primates were distinguished from other early mammals by agile and grasping hands and feet. These grasping appendages (or cheiridia, as zoologists prefer to call them) allowed nimble movement and secure footing on branches while animals searched for food and each other in forested environments of the Paleocene and Eocene epochs, some fifty to sixty million years ago.[2] All arboreal mammals rely on the sense of touch because of this need for rapid and sure-footed locomotion through trees. In primates, however, this sense has been elaborated on the hands and feet by natural selection as it worked to enhance the accuracy and precision of touch. Many anatomical specializations are responsible for the exquisite tactile sensitivity in primate fingers.[3]

The ends of primate fingers and toes are expanded into large, voluptuous digital pads that house sensory nerve endings, blood vessels, and sweat glands and are covered with fingerprints. The smooth skin of these pads is densely packed with nerve endings that permit highly sensitive tactile discrimination, finely tuned differentiation of temperature and texture, and dexterous manipulation. The digital pads contain an array of receptors that register light touch (Meissner's corpuscles), constant pressure (Merkel's corpuscles), deep pressure and vibrations (Pacinian corpuscles), temperature (Ruffini corpuscles), and pain (free nerve endings) (figure 25). These receptors detect signals from the environment, which the brain interprets as sensations; in fact, a large fraction of the cerebral cortex (the primary sensory cortex) is devoted to this task. Underneath the digital pads are small bones (the distal phalanges), which are relatively wider than those seen in other mammals. Primates evolved broad nails to physically support the enlarged and highly sensitive pads and to protect the vulnerable ends of fingers and toes.

The sensitivity of human fingertips is remarkable, and it is particularly acute in people who have lost their eyesight. Individuals who have been blind from birth or who became blind at a young age are able to detect details of texture and subtleties of form that escape the sighted. Their hands

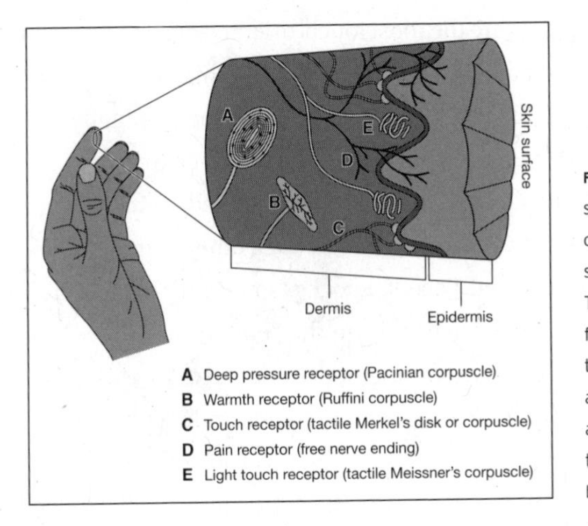

FIGURE 25. Several kinds of sensory receptors in the skin contribute to the subtlety and sensitivity of touch in primates. These receptors detect signals from the environment, which the brain interprets as sensations and the individual experiences as pressure, touch, pain, vibrations, and so forth. (Illustration by Jennifer Kane.)

become expert "observational instruments," which can discern even tiny details of a surface or protuberance. In a person with long-term blindness, the part of the brain that sighted people use to interpret visual information (the visual cortex) is actually recruited to interpret stimuli received through touch and hearing. This change accounts for the ability of people with long-term blindness to read Braille quickly.[4]

In primates, much of the cortex of the brain is focused on interpreting the signals transmitted from the sensory receptors of the hands, face, and feet. In contrast, other mammals, typified by the rat, use a significantly smaller proportion of the so-called sensory strip of the cortex for interpreting stimuli from the hands and a concomitantly larger percentage for receiving signals from their vibrissae, or whiskers. Whereas many other mammals touch and interpret their environment with the sensitive tactile vibrissae on their muzzles, primates assay theirs primarily through their hands. "Don't touch!" goes against just about every impulse in a primate's body.

In some mammals, touch receptors are used for more than touching. Bats, for example, are equipped with hair-bearing Merkel cells in their wings that allow them to detect air flow on the wing's surface.[5] When bats lose lift while in flight, the receptors signal the brain to adjust the wings' orientation to

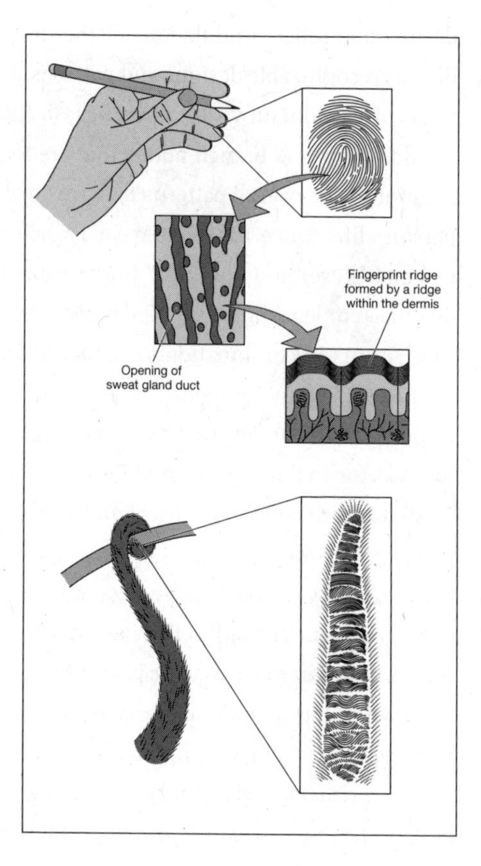

FIGURE 26. Fingerprints are formed as tiny ridges in the surface of the epidermis by undulations of the underlying dermis. The pores of sweat glands are located in the valleys between ridges. Fingerprints display unique combinations of anatomical minutiae (end points and branch points) and patterns of pores. Epidermal ridges are also found on the undersides of the tails of some large New World monkeys. (Illustration by Jennifer Kane.)

avoid stalling in midair. The extraordinary properties of these cells became clear when creative investigators removed the bat's wing hairs with depilatory cream of the kind that women use to remove hair on their legs. After this treatment, the bats could fly only in a straight line and were unable to execute the usual tricky maneuvers that allow them to avoid obstacles. After the hairs grew back, they could once again resume their acrobatics.

On our fingertips, we find one of the most widely recognized attributes of human skin. Fingerprints, known also as friction ridges or dermatoglyphics (literally, finger writing), appear on the palms of the hands and the soles of the feet of many mammals. They are tiny ridges of the epidermis that reflect undulations of the underlying dermis (figure 26) and display recognizable designs and patterns. In addition to whorls and loops, fingerprints contain visible openings for the pores of sweat glands as well as end points and branch points that are located in unique places and orientations. The overall patterns that are created never change throughout a person's life. Since 1872, when Sir Francis Galton recognized the importance of these "singularities," fingerprints have been used to identify individuals for legal purposes. Today, we can use all the anatomical details of fingerprints in combination to statistically eliminate the chance of incorrectly identifying someone.[6] Before the days of personal identification through DNA, fingerprints were the only universal and legally accepted method for making positive identifications. They remain today the most trusted and reliable biometrical tool available for this purpose.

While the forensic aspect of fingerprints is of great interest to governments and law enforcement, zoologists are more interested in the functions of fingerprints that led to their evolution in the first place. Actually, their most important role in nature is to enhance friction and to help ensure that fingers and toes placed on a branch, a slippery surface, or an object being handled will not slip off. In other words, they help to provide a secure grip. Among primates, dermatoglyphics appear not only on the palms and soles but also on the undersides of the prehensile tails of New World monkeys (figure 26), which serve as a "fifth hand" to help the monkeys secure themselves in trees while they suspend themselves from a branch or swing from one tree limb to another.

Although not fingerprints in the strict sense, the surfaces of gecko feet provide another example of skin on appendages that has evolved specialized features for gripping and locomotion (figure 27). Gecko feet contain thousands of setae, tiny filamentous projections composed of keratin that allow the animals to grip and move across smooth surfaces, even when they are upside down. This ability is based on the simple intermolecular attraction provided by van der Waals forces, which produces dry adhesion between the setae and the surface.[7] The stickiness of gecko feet and the animal's

FIGURE 27. The skin of gecko feet features dry, adhesive setae composed of keratin, which permit the animal to cling to smooth surfaces using basic principles of molecular attraction (van der Waals forces). (© Mark Moffett/Minden Pictures.)

delightful acrobatic maneuvers result not from some special chemical glue present in gecko skin, but from the surface density of the setae.

For primates, getting a grip on the environment with strong and sensitive fingers was only part of their evolutionary story. Primates were successful in their early history in large part because they were able to establish ecological niches in the widespread tropical forests of the Eocene epoch. In these rich and salubrious environments, primates evolved preferences for high-quality foods such as very young leaves and ripe fruits, which provided relatively large amounts of plant protein and readily digestible sugars. How were they able to find and select these succulent items? Comparative anatomical and paleontological evidence indicates that they evolved more elaborate senses of sight and touch. Large eyes, set relatively closely together for good depth perception and equipped with an array of color receptors, helped them locate appropriate foods from a distance. As the animals approached attractive foods, they could use both touch and smell to assess the texture, softness, and quality of the food.[8] Softer, riper fruits, for example, are richer in sugars and nutritionally superior to harder ones because they can be digested quickly and at a low energy cost to an animal. Thus the fingers we use in the supermarket today to check the ripeness of a plum or an avocado evolved their sensitivity millions of years ago, among the ancestral primates.

In addition to food selection, the need to communicate and bond was, no doubt, a driving force in the evolution of a finely tuned sense of touch among primates. Social animals like primates reinforce bonds between individuals through touch. For most mammals, this kind of communication consists of nuzzling and licking, focused on their muzzles and tongues. In primates, however, social communication emphasizes facial expressions when animals are physically separated and touching with the fingers and lips when they are close. Within social groups, closely related primates or animals that have developed friendships greet one another with reassuring facial expressions and touches. Infants attract the greatest amount of touching attention, at first from their mothers and then, in many species, from other females. Touch benefits all parties involved. Primates who are involved in greater amounts of social touching experience less stress and grow faster (if they are young). Tactile satisfaction during early development is critical for healthy behavioral development in all primates, including humans, and infants who are deprived of it develop behavioral inadequacies in later life.[9] Put simply, the absence of touch equals stress.

Besides cementing old social bonds, touching can also help to make friends. Monkeys who visit a new group are more likely to be accepted by that group if they develop consistent grooming relationships with its members.[10] Touch is reassuring because it communicates a lack of hostility and an impression of friendship. This is why caring touch is a useful, if not essential, prelude to sex in many primates, especially in humans. Touch involving the most sensitive parts of our bodies—the lips, fingers, and external genitalia—allows the physical intimacy that leads to sexual intercourse and reproduction. Evolution does not get much more fundamental than that.

During birth, primates are exposed to a sequence of intense tactile experiences: they are massaged by the muscular walls of the uterus during labor contractions and are pushed through the pelvic outlet. In humans, this process is famously prolonged, often leaving mothers completely exhausted. Fortunately, in most cultures, women in labor receive generous amounts of comforting touches from relatives or midwives. For the newborn infant, the physical rigors of the birth process appear to help prime the nervous system for functioning outside the womb, through direct physical stimulation to the autonomic nervous system that regulates circulation of the blood and respiration.[11]

Immediately after birth, most primate mothers—including human mothers in most societies in the developing world—caress and hold their infants and allow them to start suckling at the breast. For both mother and infant, the intense and pleasurable contact between the nipple and the lips, areas of skin that are densely invested with sensory receptors, initiates a cascade of important physiological changes and positive sensory feedback. Good mammalian mothers spend most of the first days after birth in close physical contact with their infants—holding, rubbing, fondling, and encouraging nursing—and similar patterns of touching are seen in most nonindustrialized human societies, where newborn infants spend their first weeks nursing and sleeping on their mother's body (even if the mother is working) or being cuddled and massaged.[12]

Much has been written about the trend in industrialized countries over the last century to reduce the amount of touching—especially mother-infant contact—during immediate postnatal life. Bottle-feeding babies, which was long touted as the modern solution to infant care and nutrition, not only is nutritionally inferior to breastfeeding but also leaves the baby deprived of the tactile stimulation necessary for normal physical and behavioral development. This problem has raised concerns about the plight of premature infants, who are physically isolated from human touch because they need a controlled atmosphere and near-constant medical intervention.[13]

Infants deprived of reassuring contact with their mothers after birth and in early infancy suffer biological and psychological stress that has ramifications throughout their lives. Among animals, laboratory rodent pups taken away from their mothers exhibit elevated levels of stress hormones, higher levels of irritability, and slower rates of growth than pups that enjoy normal rations of maternal licking. Probably the best-known studies of touch deprivation in infants were conducted during the late 1950s and early 1960s by Harry Harlow.[14] In one of a series of these classic, al-

FIGURE 28. In Harry Harlow's famous experiments on the importance of touch in infant development, the motherless infant monkey favored the warm and soft "cloth mother" even though it provided no nourishment. (Courtesy of the Harlow Primate Laboratory, University of Wisconsin, Madison.)

beit traumatizing, experiments, newborn macaque monkeys were removed from their mothers after birth and placed with inanimate surrogate mothers, one soft and warm (covered with cloth and heated by a lightbulb) and one hard and cold (constructed only of wire mesh but housing a bottle that contained milk). When presented with a choice between the nonlactating cloth mother and the milk-producing wire mother, the infant macaques spent most of their time with the soft and warm mother, leaving her only for the minimum amount of time necessary to suckle from her "lactating" wire counterpart (figure 28).[15] The infants derived more comfort from the warm and pliant cloth mother figure, even though she provided them with no nourishment.

Human infants thrive on touch. Apart from the satisfying skin-to-skin contact that infants receive as they are breastfed and held by their mothers, newborns in many cultures receive regular massages with oil, in which all parts of the body are stretched and the limbs and trunk are vigorously stroked, after which the infant is swaddled or covered.[16] Mothers, midwives, and caregivers swear by the benefits of this custom, averring that infants who receive regular massages are calmer, sleep more soundly, and experience precocious motor development. Although infant massage is not routinely practiced in most industrialized countries, it has been introduced into care regimes for premature infants, with overwhelmingly positive results as measured by rates of weight gain and progress through developmental milestones.[17] The benefits of massage are not limited to the infant. Depressed mothers who regularly massage their babies lessen their own depression and are more likely to play with their children on a regular basis. Older "grandparent" volunteers recruited for studies of infant massage therapy report similar benefits for themselves, including less depression and lower anxiety levels.[18]

Even after the infant stage, touch has measurable benefits for the young, as demonstrated by several studies of children housed in the orphanages and foundling homes that were common during the late nineteenth and early twentieth centuries in the United States, Germany, and the United Kingdom. Through the 1920s, assigning an infant to such an orphanage meant abandoning the child to almost certain death—not from lack of food or medical attention but from lack of touching and nurturing social interaction with caregivers. A widely cited study of children in two German orphanages found that youngsters housed in an institution where they were generously touched and played with by a warm and caressing matriarch grew faster than those under the care of a humorless martinet who avoided physical contact with children and often subjected her charges to verbal humiliation. Death and chronic disability caused by stress dwarfism plagued orphanages until caregivers introduced "mothering," previously considered unscientific, as part of daily infant care regimes in the 1930s and 1940s.[19]

Some American and European hospitals are now cautiously introducing massage as part of normal infant care because of the unequivocal short- and long-term benefits to the baby. Massage is also being used for some children with autism, who generally do not like to be touched because their skin can be hypersensitive. In at least some instances, these children seemed to like the deep pressure of massage, in contrast to random light touching. After massage sessions at school, they related better to their caregivers and teachers and also slept more soundly. Autistic adults have also reported the soothing and calming effects of deep pressure. In the short term, massage reduces a person's level of stress hormones; in the longer term, it encourages the secretion of growth hormone and thus contributes to rapid weight gain and motor development.[20]

After a period of intense dependence and nearly constant physical contact with its mother, a primate infant graduates to a long period of intermittent maternal contact. During this phase, babies spend varying amounts of time exploring their environment, nursing, being carried by their mothers, and being groomed by their mothers or others. For primates, grooming involves physically cleaning and touching the hair using either teeth (as prosimians such as lemurs do) or fingers (as monkeys and apes do). Grooming is a form of touch that primates enjoy throughout their lives. It is far more than the simple exercise of cleaning and ridding the hair of external parasites; rather, it is a primary social glue that holds primate groups together, beginning with the bond between mothers and infants.

In recent decades, ethologists, anthropologists, and psychologists studying primates in the wild and in captivity have explored the importance of grooming as a mode of communication, as a method of establishing and maintaining alliances, and as a part of the process of conflict resolution.[21] Some of the most significant of these studies have combined observations of how primates behave in various social situations with an assessment of how their hormonal and physiological systems relate to stress.[22] Findings indicate that grooming, much like massage, reduces levels of stress hormones in both the animal being groomed and the animal performing the grooming. Primates who participate in frequent bouts of grooming demonstrate less anxiety and depression than those who do not. In species such as the common baboon, infants are born into a hierarchy in which they inherit their social position from their mother. High-ranking infants garner more attention and more grooming from fellow group members, and they mature faster than lower-ranking ones. Infants born to low-ranking mothers receive little grooming from individuals other than the mother and experience greater physiological stress, resulting in lower rates of growth and higher rates of illness and mortality.[23]

Increased levels of stress hormones caused by touch deprivation also adversely affect the immune system. Experimental studies on captive infant macaques demonstrate that an animal's ability to produce antibodies in response to a challenge to the immune system is directly related to the amount of physical contact and grooming the infant experienced.[24] Infants separated from their mothers exhibited a suppressed immune response, whereas those receiving routine maternal contact—including normal amounts of grooming—were able to mount vigorous responses to the challenge.

Conflict resolution is an important function of grooming in adult primates. Primates, particularly monkeys and apes, are long-lived mammals who reproduce slowly and have mostly single offspring. Conflicts frequently arise in primate social groups over access to food, mates, or location or when a new animal is introduced into the troop. Evolving a way to prevent and resolve conflicts has been crucial in order to prevent decimation of a species by the serious physical injuries and deaths that could be incurred in fighting. Facial displays or gestures often allow individuals to avoid dangerous altercations in the first place. But when aggression escalates into energetic chases or physical fights, these clashes are almost always followed by interludes of reconciliation that, in most species, culminate in bouts of grooming, even when the altercation has led to serious injury.[25] These episodes of ritualized touching help to reassure individuals on both sides of the conflict and to restore the social bonds between them.

Grooming is a part of normal human life, even if we don't label it as such. Casual observation of people at home or in relaxed social settings reveals that most of us "groom" one another regularly through touches, caresses, and reassuring pats and rubs. In many social contexts today, especially the workplace, physical touching between unrelated people is strongly discouraged, but the need for social touch remains. The popularity of office massage, head massages in hair salons, and the bewildering array of wraps, massages, and other spa treatments now available in many industrialized countries speaks of humans' abiding desire to be touched, even if we prefer to discuss the experiences only in terms of improving our health or appearance.

People in all cultures touch one another. But touching, which is a learned behavior in humans, isn't necessarily the same from one culture to another; standards of permissible and desirable touching, both in public and in private, can vary widely. Different styles of touching abound (think of all the different kinds of handshakes and hugs you've experienced), and they are often used to send social signals. For example, embarrassing confusion results when someone kisses a hand that has been offered for shaking, when a handshake resembles an encounter with a dead fish or a vise, or when one is unexpectedly enveloped by a smothering embrace and liberal kisses. In many cases, only certain people can engage in certain kinds of touching—like a doctor performing a medical exam. When it comes to humans touching other humans, context is everything.[26]

In his book *Touching,* anthropologist Ashley Montagu reviewed the literature on cultures of touching and found stark, pervasive differences between "contact" and "noncontact" societies. These patterns are established from birth. In highly touch-oriented cultures, infants are liberally held, fondled, and massaged by mothers and other caregivers. These cultures disapprove of cradles, strollers, and other devices that physically isolate babies from regular touch; people who want to adopt such paraphernalia as part of a "modern lifestyle" are ridiculed and chastised.[27] In touch-averse cultures, by contrast, babies are often deprived of maternal and human contact except during small fractions of the day when such behavior is socially sanctioned. As the children in these societies mature, they are increasingly physically distanced from their parents and often further deprived of physical contact, despite their entreaties.

Modern American culture is mostly touch-averse, especially in the settings of schools, hospitals, nursing homes, and most workplaces, where concerns about litigation have reduced the amounts of acceptable touching to a minimum. Children growing up in this culture adapt by learning to express their emotions through words and facial expressions rather than by touch. But such accommodation comes at a cost, as older children and adults suffer from awkwardness in demonstrating physical affection and ineptness in body relationships with others in general.[28] From the point of view of comparative primatology, touch-averse cultures are an anomaly, and the bouts of depression, anxiety, and more serious forms of social pathology among individuals who live in them are entirely predictable.

Touch is not only a sign of friendship—it can take the form of physical contact inspired by anger or aggression. Anger directed against children can be manifested in physically hurtful forms of touching such as slapping, spanking, and caning, as adults substitute painful stimulation of the skin for comforting touch in order to show displeasure and mete out punishment. Primate mothers use slapping and cuffing very occasionally but highly effectively when youngsters pester them—for example, when a mother wants to wean her offspring, and the infant continues to beseech her for milk and tactile attention. Primate mothers are generally exceedingly patient under such circumstances, and physical punishment is rare. In humans, however, corporal punishment is not rare. Ironically, the physical victimization of children, women, and the elderly is rife in touch-averse cultures. Children in such circumstances are easy targets for physical aggression because of their small size and relative helplessness.

Children deprived of nurturing contact but subjected to routine physical punishment are prone to serious behavioral disturbances, drug addiction, and physical violence.[29] In extreme examples in recent history, cultures of beatings and canings arose in order to punish children for their misdeeds and, purportedly, to toughen them for life in a cruel and competitive world. These children (mostly boys), in otherwise noncontact cultures, were beaten at home and in school, with constant warnings and threats. Children treated this way repeatedly learned to dissociate the infliction of pain from a show of emotion, a process that enabled them to inflict hideous pain and torture on others.[30]

In light of the primacy of touch in promoting human well-being, the value of interpersonal physical contact during sickness and old age should be evident. Healers throughout time and in most cultures have used the "laying on of hands" as an essential part of caring for the sick.[31] The manifold physiological benefits of touch are measurable: pain relief for cancer and arthritis sufferers, gains in peak air flow for asthmatics, increased lymphocyte counts for those with HIV or AIDS.[32] These benefits derive from a succession of responses that begin when tactile and pressure receptors in the skin are stimulated. This activity in turn stimulates the central nervous system, prompting the secretion of endogenous opiates (endorphins and related compounds that make us feel good). The result is both pleasure and relief, with measurable reduction of anxiety and stress levels and strengthening of the immune system, whose functioning is inhibited by high stress levels.

Touch benefits not only those who are sick and in obvious physical pain but also the elderly. In industrialized societies, older people often live in isolation or in care facilities and are thus deprived of physical contact. The benefits of touch include feelings of improved well-being and fewer signs of senility, such as irritability, forgetfulness, irregular eating, or careless grooming. In one study, nursing home residents who received regular massages, hugs, and thoughtful squeezes acted younger, demonstrating greater alertness, vitality, and better humor, than those who received no such contact.[33] The mere act of touching another living being is beneficial—witness the success of programs that encourage pet ownership among the elderly and "grandparent" volunteers who hold and massage premature babies in hospitals. As good primates, humans benefit from grooming and touch—in whatever cultural manifestation it is presented—at all ages.

8
emotions, sex, and skin

Some of the day-to-day changes in our skin are slow and invisible: old skin is replaced, melanin is produced, vitamin D is formed. Other changes, however, particularly those that reflect our emotional state, are sudden, highly visible, and palpable. Our skin often "thinks" before we do. It can react to a stimulus, leaving us with goosebumps, sweaty palms, or red faces, even before we can identify the cause.

Human skin contains a vast network of nerves, including sympathetic nerve fibers, which belong to the autonomic nervous system.[1] The job of this system is to maintain the body's internal environment by controlling such "automatic" functions as breathing, circulation, and digestion. In particular, the sympathetic nerves are responsible for the "fight or flight" reaction in the face of stress. When we experience fear, they respond quickly and dramatically, increasing the heart's output and diverting blood from our skin and gut to our skeletal muscles. By rapidly channeling the body's resources to those organs necessary for confrontation or escape, the sympathetic nervous system literally can "save our skin." As these nerve fibers in the skin fire, they simultaneously help to constrict the small arteries in the dermis, activate the sweat glands, and stimulate the tiny smooth muscles in our hair follicles to produce piloerection. The result: pale, clammy skin, hair standing on end, goosebumps. Add dilated pupils, and you have the face of fear. This is an extreme reaction, which the body launches only in an emergency.

Less extreme responses are evident when we are anxious or excited rather than terrified. At these times, the sympathetic nervous system is stimulated to a lesser extent, and the effects are what we call a "cold sweat"—icy, damp hands and unpleasantly sweaty armpits. This phenomenon has its own ancient evolutionary history. The eccrine sweat glands of the palms and soles are, evolutionarily, the oldest sweat glands in our bodies. In contrast to those found in the rest of the body, these sweat glands respond to emotional, rather than thermal, stimulation, which is delivered by the sympathetic nervous system.[2] (The eccrine glands of the face, armpit, and groin respond to both types of stimuli.) The cold sweat that you experience when you feel nervous or agitated is technically referred to as emotional sweating, to distinguish it from the thermal sweating that helps to keep you cool. Although little research has been done on the origin of this reaction, we might reasonably speculate that stimulation of the sweat glands on the hands and feet of ancient tree-dwelling primates made for more secure locomotion, especially when an animal was being chased. In small primates, moisture added to the subtle texture of fingerprints would have allowed a firmer grip on branches and tree trunks, and a slight dampness on the hands and feet would heighten skin sensations, making it easier for an animal to "read" its environment through its fingers.

Emotional sweating has played a significant role in the arena of forensic science. This type of sweating betrays the mood state of nervous anxiety, even at its most subtle level. Scientists can measure the intensity of this reaction by looking at changes in the skin's ability to conduct electricity—an indication that the skin has become moister as a result of sweat gland activity. Measuring electrodermal responses forms the basis for the con-

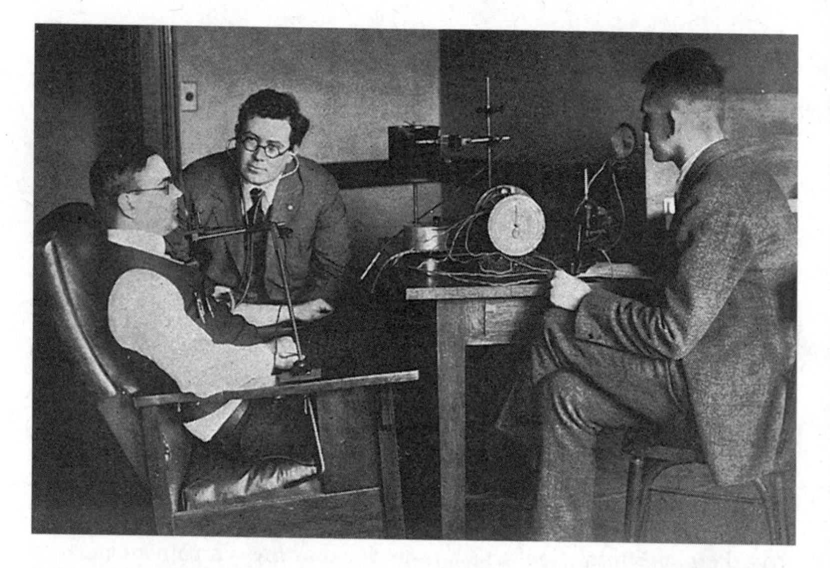

FIGURE 29. For decades, law enforcement authorities routinely used the polygraph to assess the likelihood that a person suspected of a crime was deceiving the police. The polygraph measures minor changes in skin conductance that result from nervous anxiety and the consequent emotional sweating. The lie detection device pictured here was a predecessor of the polygraph. (Photograph 1920. Robert C. Givler. *Psychology: The Science of Human Behavior*. Harper.)

troversial polygraph, or lie detector, test, which measures minor changes in skin conductance (figure 29). For many years, the criminal justice system in the United States utilized the polygraph test to help determine whether a person suspected of a crime was lying. In recent decades, experts have questioned the reliability of the polygraph because skin conductance can change not only from deceptiveness but also from the fear elicited by taking the test. Although the results of standard polygraph tests are no longer admissible as evidence in American, Canadian, or Israeli courts, the polygraph is still used in some contexts—by employers, for example, or by people accused of crimes who voluntarily submit to testing in order to help establish their innocence.[3]

In addition to sweating, the color of our skin can also mirror our emotional state. Human skin is drained of color when we are frightened, but it can turn visibly red when we are angry or embarrassed. Flushing in response to anger and blushing in response to embarrassment are fascinating phenomena, though not well understood. These reactions are generally confined to the face, especially the cheeks, forehead, and ears, because the arteries in these areas have an enhanced ability to dilate, or widen, and thus increase blood flow. The intensity of these responses differs from one person to another, in part because people handle stress differently.[4] Some people, for example, blush deeply with little emotional provocation, whereas others barely blush at all even when the situation warrants it.

The reactions of facial skin during anger have been studied more thoroughly than those involved in blushing. "Turning red in the face" is synonymous with the most forceful expressions of anger in many cultures. The whole-body physiological effects that accompany such anger are more extreme than the body's response to any other emotion.[5] These effects build incrementally, as heart rate and blood pressure rise, peripheral arteries constrict, and the emotional sweating reaction kicks into gear. As anger builds, branches of the external carotid artery undergo a dramatic vasodilation, or increase in the artery's diameter, increasing blood flow to the face. It is at this point that the face turns red and the main trunk of the artery visibly stands out on the side of the neck. Such rage is an extraordinarily unpleasant experience, whether you are feeling it yourself or observing it in others.

For years, the reasons for the "flush of anger" were not well understood. Why, after all, should the face turn red when a person is preparing for a fight, when the body's sympathetic nerves are working to increase cardiac output and constrict most of the blood vessels in the skin in order to divert more blood flow to skeletal muscles?[6] It seems contradictory that at the same time the cutaneous blood vessels in one area would dilate. The possible reasons for this seeming contradiction are intriguing. One explanation notes that increased facial blood flow during extreme anger appears to serve as a safety valve when blood pressure rises to extremely high levels.[7] When the pressure receptors in the arteries of the neck detect dangerously high levels of blood pressure, the walls of these arteries actually relax, permitting blood to surge into the face and thus helping to slow down increases in heart rate and blood pressure.

Another explanation emphasizes the visibility of an angry red face. Reddening of the face during extreme anger is a highly conspicuous sign of a powerful negative emotion, and it serves as an unambiguous warning to "stay away or else." In evolutionary terms, it is an obvious signal warning potential combatants that a physical attack is likely or imminent. It seems plausible that natural selection may have favored accentuation of the anger flush because of its success as a way to avoid conflict, an ability that has been consistently important in the evolution of monkeys, apes, and—especially—humans.

Blushing is similar in some ways to the flush of anger: the face reddens when the sympathetic nerves in the face that govern vasodilation are stimulated and blood flow to the face increases. People blush when they are embarrassed or receive undesired social attention. It is a graded response, sometimes involving only the cheeks and ears, but it can spread to the forehead and neck if a social situation is sufficiently uncomfortable.[8] A complete suffusion of the face ("turning beet red") is uncommon, but for those afflicted by it, this intense blushing can itself be a cause of anxiety. Some people who are prone to profuse blushing go to great lengths to avoid embarrassing situations; such blushing anxiety is, in fact, a recognized condition.[9]

Our skin registers reactions not only to the clouds of embarrassment or rage that pass through our lives but also to longer-term stress. Chronic stress and depression have adverse effects on health and life expectancy and alter the body's immunity in ways that render it more susceptible to diseases, including those of the skin. Common skin diseases such as eczema (various rashes), psoriasis, and hives are made worse by chronic stress. When an individual suffering from such a condition feels stigmatized because of the way the skin looks, the stress may become more severe, causing a heartrending cycle of increased anxiety and withdrawal. This fundamental connection between the skin's appearance and chronic stress is now widely appreciated in medical circles.

Although the literature on human evolution has paid little attention to the subject of facial reddening, we can reasonably speculate that reddening associated with both extreme anger and embarrassment evolved early in human history. The development of bipedalism led to fundamental changes in modes and patterns of social signaling. We know, for instance, that male canine teeth became smaller, probably because they were less important in social displays and actual fighting.[10] Possibly at the same time, humans also began to lose their body hair and, concomitantly, their ability to visibly demonstrate agitation through piloerection (refer back to figure 4). These changes would have placed increased emphasis on body posture and facial expressions as the most highly visible signals of mood states, especially negative ones. Facial muscles would have become more elaborate in order to produce more nuanced facial expressions. Today, we find that all humans, regardless of skin color, experience facial reddening when enraged or embarrassed; people with darkly pigmented skin "turn red in the face" physiologically even though the reddening isn't as noticeable as in those with light skin. This shared capability indicates that the evolution of facial reddening preceded the evolution of enhanced melanin pigmentation in the human lineage and that ancient hominids predating early *Homo* were capable of such displays.

Skin, in subtle and not-so-subtle ways, also reflects our existence as sexual beings. The same is true for our primate relatives, who tend to fall into the not-so-subtle category. In some species of Old World primates—specifically, macaque monkeys, baboons, and chimpanzees—the rumps of sexually mature females turn pink (even bright red) and become hugely swollen when the females are sexually receptive. Pink females always seem to attract the attentions of males, often prompting copulation right there and then. The reasons for this sexual connection are fascinating.

Primates are highly visual animals, and easily visible cues are more effective attention-getters than smell alone. As many female primates approach ovulation, the perineal skin around the external genitalia reddens, becoming most livid and swollen during the few days when they are actually ovulating. In macaques, baboons, and chimpanzees, a female's perineal skin is suffused with blood vessels that respond to cyclical changes in hormone levels.[11] The color change that occurs in the skin is the result of increased blood flow, and the swelling is caused by an extravagant accumulation of extracellular fluid, which subsides quickly after ovulation. In some females, the swelling, at its maximum, adds as much as 25 percent of the animal's body weight. This display serves, quite literally, as a red flag advertising sexual availability and fertility.

For years, primatologists pondered the question of why some females have more swelling than others. It turns out that, in this instance, bigger is genuinely better, in terms of reproductive success and evolution. A long-term study of wild baboons has shown that females with greater swelling attained sexual maturity earlier than those with smaller swelling. Further, the females with the puffiest bottoms produced more offspring during their lives and had more offspring survive on average than did females with less puffy bottoms. Male baboons found females with larger swellings considerably more sexually attractive and made more of an effort to keep them as partners when the females were in season. The males consorting with the puffiest females spent more time defending themselves against the aggressive advances of competing males and more time grooming the females in order to ensure their willingness to copulate.[12] Large sexual swellings are an example of what zoologists refer to as an "honest signal," a reliable indication of evolutionary value that is physiologically costly to produce but that yields results in terms of enhanced reproductive success. Thus, big pink bottoms are hot stuff—dependable signs of a female's reproductive receptivity and lifetime reproductive potential.

There is also evidence that, at least in chimpanzees, obvious sexual swellings inhibit aggression that might otherwise be directed toward sexually receptive females. When they reach sexual maturity, female chimps leave the group in which they were born and transfer to a new group. This passage is fraught with risk because chimpanzees are wary of outsiders, and they will physically assault and sometimes kill trespassers and interlopers.[13] Female chimpanzees almost always stage their transfer to a new group when their sexual swellings are at their biggest and brightest, which often allows them to attract the attention of a male in the new group and begin a consortship with him. Through this association, the female can sometimes slip relatively unobtrusively into the new group under the male's (perhaps temporary) protection. Even without an established consortship, pink females are less likely to be attacked because males will be interested in them more as potential sexual partners than as targets of aggression.

A flashy pink bottom does not mean that the female will have an easy time becoming established in the group, however. After the "first flush" is over and her bottom once more becomes pale and flaccid, an immigrant female must work to insinuate herself into the group's social structure, a lengthy process that requires her to befriend and ingratiate herself with the adult females who form the core of the group's power structure. She may also be involved in aggressive and often brutal physical encounters with resident females. In at least one recorded instance, a female's red flag did not prove to be a free ticket for entry into a new group, but instead made her the target of a vicious attack, in which she was bitten and her bottom was slashed open by hostile residents.[14] This appears to be an exceptional case in which the "honest signal" of sexual availability marked the female as a target for attack. It is rare in nature for such signals to be ignored.

Among humans, skin is also intimately involved in sexual activity. When some people are sexually aroused, their skin becomes flushed, usually around the neck and chest. This sexual flushing of the skin often persists for several minutes, lasting longer than blushing, and leaves the person "hot and bothered." This phenomenon has not been well studied, but it appears to be the result of vasocongestion, that is, a slowing of blood flow, in the small veins of the skin on the neck and chest.[15]

The skin is the largest sexual organ of the human body, although we don't usually think of it as such. Much of the pleasure of sexual intimacy comes from the exquisite expectation of touch and the delight and relief of skin-to-skin contact with another person, before, during, and after the sex act itself. Certain parts of the body are especially sensitive to sexual touch; this heightened sensitivity may result from a greater density of nerve networks closer to the surface of the skin. Caressing, stroking, or lightly brushing these erogenous zones increases sexual pleasure and arousal. In addition to the genital areas, erogenous zones can include the breasts and nipples, the neck, the inner thighs, and the mouth, although this varies from one person to another. The scientific literature on the biology of erogenous zones and the reactions of skin during sex is disappointingly but perhaps not surprisingly sparse. Fortunately, this is one area of science in which most people can boast some expertise, gained from personal experience.

Our skin also reacts when we're thinking about sex or talking about it. Many such reactions are socially conditioned: when a culture treats sex as a sensitive or taboo subject, some people have a difficult time talking about it, and their faces turn red with embarrassment when they are forced to do so. They even turn red when they see animals having sex.

Zoos are great places to observe animals, and nonhuman primates are some of the most popular attractions for kids. As an anthropologist, I've spent many hours watching monkeys and chimpanzees in zoos—and watching them watch us. I've also enjoyed watching other people watch the primates. Because it's actually rare to observe a group of monkeys or chimps when there isn't something sexual going on with the animals, I've often noticed the following scene unfold. A child asks an adult what those two monkeys are doing, and why that one has such a funny-looking pink bottom. The adult stands there, developing a red face and fumbling for something to say, while pondering the pillow of livid tissue on the female monkey's backside. Some people appear to become quite undone by this, even if they know the answer to the child's question. The only thing that's clear is that they wish that both they and the monkeys were somewhere else. Pink bottoms involve both ends, as it were, of the question of sex and skin.

9
wear and tear

As the body's protective covering and screen, and its first line of defense against the environment, the skin has evolved to withstand a barrage of sharp, nasty, biting, and corrosive insults. But it is not perfect: it ages, scars, and suffers the ravages of disease and environmental assaults. Most of the ills that befall the skin are temporary and reversible, but some are lasting and cumulative and eventually take a toll on both its appearance and its functioning.

Since antiquity, people have documented and contemplated the various conditions and problems that affect the skin. For thousands of years before the birth of modern medicine in the eighteenth century, the skin alone testified to the state of a person's health, displaying most of the known signs and symptoms of disease.[1] As a result of this long history and the development of a branch of Western medicine devoted exclusively to the skin—dermatology—the catalog of recognized skin conditions is bewilderingly large. In addition to being overwhelming, perusing a comprehensive textbook of dermatology can also be unpleasant. We can immediately relate to diseases of the skin, more easily than to, say, diseases of the liver. Looking at a picture of a skin disease, we can imagine the distress and self-consciousness the sufferer must feel. We empathize, and, if the condition depicted in the book is sufficiently grim, we feel physical discomfort ourselves. Our skin crawls.

This chapter's goal is not to induce this level of discomfort, but rather to examine some of the more typical "skin events" that we might encounter in the course of our lives, in roughly chronological order. Some we are born with—birthmarks, for example. Others, including scars, pimples, and wrinkles, happen to all of us and are simply the consequence of being alive on the planet for any length of time. Still others, such as warts, burns, and skin cancer, are medical conditions sufficiently widespread that they deserve our attention.

Birthmarks and Moles

Because the surface of the skin is large and subject to diverse genetic and environmental influences during prenatal development, infants are often born with birthmarks. Birthmarks occur for various reasons; many are caused by small errors in the development of blood vessels in the womb.

One of the most frequently seen birthmarks is called the fading macular stain, also known by the old names salmon patch, stork bite, or angel's kiss. This minor anomaly in blood vessel formation usually occurs on the face and neck and disappears in the first year of life.

Another common birthmark appears as a small, blue-black patch on the lower back. It is referred to as the Mongolian spot, because of its frequent occurrence among people of East Asian descent. This spot represents a collection of dermal melanocytes, pigment-producing immigrant cells that were somehow stalled as they attempted to migrate to the epidermis during early embryonic development.

The unfortunately named port-wine stain is one of the most well-known birthmarks, caused by malformed and dilated capillaries in the skin. The port-wine stain, which is permanent, tends to deepen in color over time, from pink to dark red or violet. Although they are not a risk to health, these sometimes large and livid marks that occur mostly on the face can make individuals highly self-conscious about their appearance. Since the 1980s, pulsed-dye laser therapy, which can significantly lighten the mark with a low risk of scarring, has revolutionized the treatment of these birthmarks.[2]

Nearly all adults have one or more moles, the vast majority of which are harmless aggregations of melanocytes.[3] Most moles occur on parts of the body that have been exposed to the sun at some time, and many types are linked to changes in cellular physiology caused by exposure to UVR. On very rare occasions, these moles can develop into melanoma, a dangerous form of skin cancer. Doctors advise everyone with moles, especially fair-skinned individuals, to check their skin periodically for changes in the size, thickness, or color of their moles.

In the centuries before modern medicine offered a scientific look at these anomalies, birthmarks and moles were believed to have deep meanings. Birthmarks, many thought, could disclose what had influenced a mother during her pregnancy, while moles could foretell an individual's fate later in life. Portions of many treatises and manuals from seventeenth- and early eighteenth-century Europe record the likely meanings and consequences of such marks. Events during a pregnancy, it was believed, contributed to the color and shape of particular birthmarks or more generally to the look of a child's skin. For example, a newborn's port-wine stain birthmark might be attributed to someone who had spilled port on the mother months earlier. A mother frightened by an ape or a bear might produce a hairy child, while a woman terrified by a fish might give birth to a baby with scaly-looking skin.[4] Although these interpretations may strike us as bizarre, we might consider how some of our own "explanations" will be viewed in the future.

Scabs

For most people, scabs are one of the first unpleasant things they see on the surface of their own body. When the skin is scraped deeply enough to penetrate the epidermis, the blood vessels in the dermis can be damaged and start bleeding. Platelets in the blood come into contact with collagen and other components of the skin that have been exposed by the injury. This contact induces the platelets to release clotting factors and other substances in order to stop the bleeding. As the bleeding stops, the healing has already started.[5] Special white blood cells called neutrophils then arrive at the site to begin the process of ridding the area of foreign materials, unwanted microorganisms, and damaged tissues. This process is aided by the development of local inflammation, fueled by cytokines, molecules that coordinate subsequent healing.

Once the wound site has been cleaned out, fibroblasts migrate to the area to start laying down new collagen on the scaffolding of the original clot. The chemical activity occurring at this time is intense, as the fibroblasts at the wound site mature and produce new proteins to speed healing. In order to restore the dermis, different types of collagen are produced at the wound for the next several days. The collagen undergoes continuous remodeling to physically fill in the injured region. Simultaneously, new blood vessels become established in the area. While all this is going on, the epidermis is preparing to repave its surface by moving new keratinocytes to the location of the injury.

The scab we see on the surface of the skin is, in part, the debris from the construction site, because it contains elements of the original blood clot and other emergency proteins that entered the wound in the early stages of healing. The scab is also the skin's natural bandage, because its thickness protects the underlying tissues during healing. Over the course of a few days, scabs change in their composition and appearance: at first, they are thin, somewhat flexible, and reddish-brown; later they become thicker, harder, and darker. Just before a scab falls off, it becomes smaller in diameter and more prominent, as its collagen fibers contract and the underlying skin becomes fully resurfaced. There isn't a child alive who hasn't picked at a scab because it was itchy or ugly, or a parent alive who hasn't told a child not to do it. The old refrain, "It will only get worse if you do that," is correct because the presence of a stable scab ensures the formation of smooth new skin underneath, reducing the likelihood of a lasting scar.

Scars

When the skin is cut, burned, or bitten, a scar forms when collagen is laid down in the wound as part of the healing process. Some scars are barely visible; others are highly noticeable. The appearance of a scar depends on the nature of the original injury to the skin. Clean cuts like those made by a knife or razor heal quickly, often with almost imperceptible scars because the edges of the wound remain close together during healing. Jagged cuts and serious burns heal with more prominent scars because the edges of the wound are farther apart as it heals. Scars can also result from pimples, acne, or systemic infections involving the skin, such as chicken pox or smallpox. These scars often resemble small pits or depressions and are known as atrophic scars (figure 30).

The formation of scars is usually uneventful, but occasionally an overgrowth of collagen can result in an abnormally prominent scar. When scar tissue builds up within the confines of the original wound, it creates a raised, or hypertrophic, scar. When collagen overgrows the boundaries of the original wound and scar tissue continues to form, piling up thick, hard, and unsightly accumulations of collagen fiber, the resulting raised and sometimes tender scar is called a keloid.[6]

Hypertrophic scars and keloids are noticeable and can cause people to feel self-conscious when they occur on areas of skin that are routinely exposed, such as the face. They also lack many of the normal properties of skin—they are hairless and have no sweat glands or natural elasticity. The more collagen fibers that are laid down during the healing process, the stronger and less flexible the scar will be. Keloid scars develop more often in people with moderately to deeply pigmented skin, for reasons we do not yet understand. This phenomenon has contributed to the traditions of decorative scarring in many parts of Africa, as we'll see in the next chapter.

Like everything else in the body, scars and scar tissues change with age.

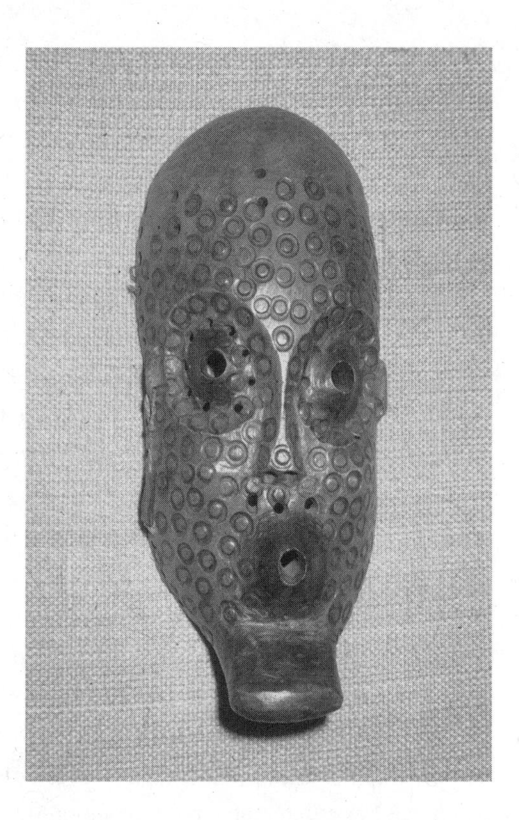

FIGURE 30. This mask from the southwestern African country of Angola depicts the deep pitted scars that characterized the once widespread disease of smallpox. (Courtesy of Edward S. Ross.)

Most scars undergo a process of "normalization" over time, becoming less prominent as the collagen networks within them are remodeled and as temporary blood supplies that were established to heal wounds naturally regress.

Bites and Stings

Like other warm-blooded creatures, humans are susceptible to being bitten by arthropods, members of the large group of invertebrates that includes insects and spiders. Usually only an annoyance, arthropod bites can involve more serious problems, depending on the organism doing the biting. Many of the diseases that exact the highest toll on human health, including malaria, sleeping sickness, dengue fever, and plague, are transmitted by insect bites. Animals are not bitten randomly: many arthropods are fussy about their choice of victims and exhibit a high level of what is referred to as host specificity. The coordinated evolution of biters and the bitten is one of the most interesting sagas in evolutionary biology and is now the subject of increasing basic and clinical research efforts.[7] Of greatest interest to us here is how arthropod bites occur in the first place and why they are such a nuisance.

Biting arthropods are hungry arthropods. They seek humans and other warm-blooded prey to obtain blood meals, which female arthropods require for reproduction. Insects and spiders are attracted to humans because of our warmth and scent; some are also attracted by the colors we wear and the carbon dioxide we exhale. The biting process varies somewhat from one group of arthropods to another, but it usually involves the use of serrated jaws that disrupt the skin. In the case of a mosquito bite, this small indignity is followed by the female mosquito inserting a feeding tube through which she withdraws a nourishing meal of blood, much as one would suck a thick milkshake through a straw. To ensure that her straw doesn't get clogged, the insect injects an irritating solution into the skin to prevent the blood from coagulating. This solution causes the swelling and itching associated with the bite. Scratching the area immediately around the bite spreads the anticoagulant—and the discomfort—around.

When spiders bite, they inject venom into the victim through the jaws, without withdrawing a blood meal. Spiders are generally timid creatures and bite warm vertebrates only when they feel threatened. They normally reserve their biting for insect prey, which they immobilize and consume in their webs. Biting a vertebrate yields no direct dietary reward to the insect but presumably contributes to its self-preservation. Most spider bites produce only transient pain and swelling, but those of the brown recluse and black widow spiders are more serious because the venom is more highly toxic.

Bees and wasps take a different approach. Instead of using jaws to cut into the skin, the female insect stings its victim with a modified ovipositor, the apparatus that it uses to deposit its eggs. Such stings are often painful at the time and produce an intense local reaction, which includes swelling and reddening of the skin, followed by itching. Stings deter animals like humans from approaching hives and nests, where the insects store food and raise their young. Multiple stings delivered at the same time can produce a dangerous allergic reaction, hence the universal human caution when it comes to approaching bee and wasp nests.

Scores of other arthropods, including fleas, ticks, mites, and bedbugs, bite humans when they have the chance. But only one insect has been sufficiently vexatious to humans throughout history to warrant its own damning epithet. Thanks to the louse, we have the evocative word "lousy," a synonym for all that is nasty, unclean, and rotten. Lice and humans have had a long and interesting partnership.

Modern humans commonly become infested with not one but two types of lice. Head lice *(Pediculus capitis)* appear to have the longest history of association. Found mostly on the heads of children, these insects are the bane of daycare centers and kindergartens because they can be transmitted quickly from one individual to another through close physical contact and through sharing combs and hairbrushes. Head lice deposit their small oval egg capsules (nits) near the base of hairs on the scalp. The egg is composed of a waxy covering with a breathing hole that protects the developing larva. The larva normally incubates in the warmth of the scalp for seven to nine days before exiting its egg case and searching for its first meal. If the young louse doesn't obtain a blood meal from its human host in the first twenty-four hours of its life, it will die. Severe and long-term infestations of head lice thus involve egg-laying adult lice, larva-containing egg cases, and hungry nymphs that bite the head that feeds. Although many people today use special chemical shampoos to rid themselves or their children of lice, grooming—specifically, nit-picking—was the preferred mode of delousing for most of our history.

Until recently, head lice and body lice were thought to belong to the same species, but recent detailed study suggests otherwise. Body lice *(Pediculus corporis)* behave differently, preferring to lay their eggs on clothing rather than on the body, which seems to be a relatively new affectation. Recent anatomical and molecular studies have shown that body lice are the evolutionary descendants of head lice, having created a new ecological niche for themselves in human clothing during the past forty thousand years or so.[8] Body lice may be more of a problem, as they can act as carriers for typhus and other diseases. Human behavior has been responsible for the evolution of many new organisms in and on our own bodies, and these parasites are just one example.

Burns

Burns result when the skin comes in contact with intense heat or fire. Among children, the most serious burns are usually caused by hot liquids; among adults, the most serious injuries result from contact with open flames at home or at work.[9] In partial-thickness burns, the most common type, we see redness (a first-degree burn) or a fluid-filled vesicle (a second-degree burn) at the burn site. Far more serious are full-thickness, or third-degree, burns, in which the entire thickness of the skin, including the epidermis and the dermis, is destroyed. Third-degree burns that cover large areas of skin (roughly, greater than 15 percent of the body's surface) are extremely serious injuries: they lead to the relatively rapid loss of body fluid, and these burn sites can be rapidly colonized by bacteria that can cause major infections.

When a serious burn occurs, the protective role of the skin comes into sharp focus. Treating such a burn involves many steps, from acute care to surgery and then on to physical and psychological therapy.[10] As burns heal, the damaged skin is naturally replaced by scar tissue, which forms as collagen is laid down at the burn site. The collagen then contracts as the body slowly tries to bring the edges of the wound together. With large burns, this natural process can have disastrous results because it leads to the formation of disfiguring contractures that can immobilize parts of the body that are normally highly mobile, such as the neck. Burn victims must often endure repeated surgeries in order to remobilize parts of their bodies that have become shackled by scar tissue. Scar tissue is also quite different from normal skin. Because it lacks the skin's elasticity and laminated structure, it can't stretch as a person moves. It also cannot sweat, being bereft of sweat glands and hair. It is no wonder that the search for "replacement skin" has been one of the most fervent and difficult missions in the history of medicine.

Burn treatment has advanced remarkably in the past twenty years. We have learned to better control infections, and the introduction of "artificial skin" has greatly sped up wound healing. Dedicated burn units have been established throughout the industrialized world. In the developing world, however, burns remain a major cause of death and loss of livelihood.

Dermatitis

Human skin is designed to resist all manner of chemical and biological attacks. The skin's immune system resembles a modern army, with several specialized types of defensive cells and chemicals that are deployed in different arrays and at different times depending on the nature of the attack (figure 31). The skin responds first by immediately producing chemicals that cause inflammation or fight infection. This so-called innate response then triggers an adaptive response, which is initiated by the Langerhans cells of the immune system that are based in the epidermis. When sufficiently stimulated, the Langerhans cells turn into a marauding army that can migrate from the skin to nearby lymph nodes, where they can induce the formation of infection-fighting lymphocytes. Often, this battle against foreign chemicals or pathogens becomes visible and palpable. Our skin gets red and itchy. Sometimes it develops small bumps filled with clear fluid. If the irritation lasts long enough, the skin can become thick and scaly. All of these effects are related to dermatitis.

Simply defined, dermatitis is inflammation of the skin. Because inflammation can result either from an external irritant or from a chemical imbalance within the body, dermatitis encompasses an enormous range of skin diseases.[11] The most common types, referred to as contact dermatitis, are caused by contact with a chemical irritant or microorganism. A classic example of contact dermatitis is poison ivy or poison oak. The itching (pruritis) that almost always occurs with dermatitis results from the chemicals

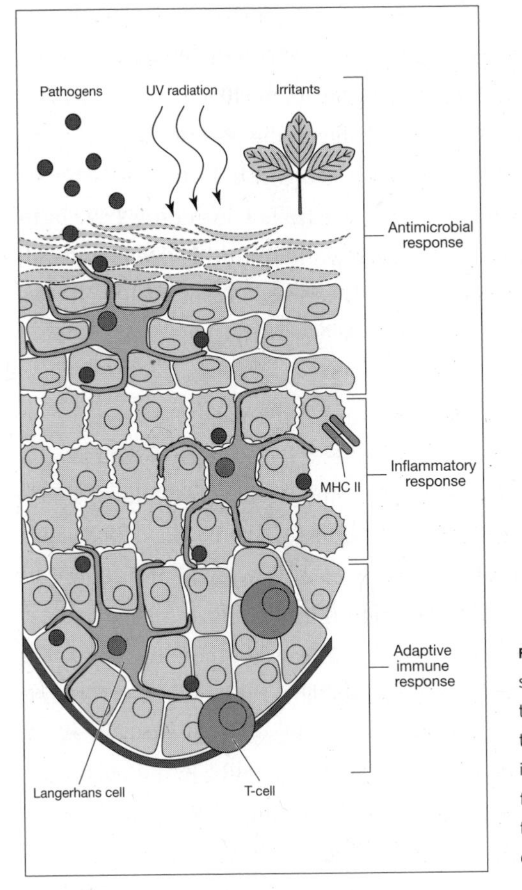

FIGURE 31. The skin's immune system operates at different levels to defend the skin and underlying tissues. This involves an immediate innate, or antimicrobial, response followed by an adaptive response that is mediated by the Langerhans cells. (Illustration by Jennifer Kane.)

the body produces as part of the inflammatory process. These in turn stimulate an itch-specific neuronal pathway. The delightful, though short-lived, relief provided by scratching is apparently related to the mild pain it causes, which inhibits those neurons.[12]

Pimples and Acne

Teenagers experience a torrent of physical and physiological changes as their bodies make the transition from childhood to adulthood, and some of these are manifested in their skin. The skin reacts to many hormones, including the powerful steroid testosterone, a sex hormone. As testosterone levels ramp up in teens of both sexes, the skin responds by secreting more sebum—an oily mixture of fats and waxes—from its sebaceous glands in the dermis.

Most of the time, the cells lining the ducts of these glands are shed as sebum is produced. But when large amounts of sebum are secreted, the duct can become blocked as the lining cells and sebum accumulate and form a plug, or comedo. A plug that enlarges and comes out of the duct is known as an open comedo, or blackhead. If it stays below the surface of the skin, it is called a whitehead. An inflamed whitehead is a pimple. Many skin care products specifically state that they are "noncomedogenic," that is, they do not lead to blocked pores and pimples. When many pimples develop simultaneously, the condition is referred to as acne. In its most disfiguring expression, inflammatory acne can produce cysts and deep, pitted scars on the face, neck, back, and chest. Acne tends to be more severe in boys because the rush of testosterone production in adolescence, which is stronger in males, stimulates the activity of sebaceous glands.

Pimples and acne are not the exclusive province of youth. Adults can also experience pimples and acne because of changing hormone levels. This "persisting acne" can be as emotionally traumatic as the adolescent variety because the visible signs take longer to disappear, as the body's inflammatory reactions and healing processes slow down with advancing age. Acne is not normally considered a life-threatening disease, but the anxiety and depression it engenders in teens and adults have sometimes led individuals to contemplate suicide. Because so much of a person's self-image is connected with the appearance of his or her skin, we should never dismiss the psychological effects of acne as trivial.

Warts

The belief that warts are caused by handling toads dates from the same era that promoted belief in the fortune-telling powers of moles. Warts in fact are caused by a family of viruses known as papillomaviruses and can be transmitted by contact from one person to another. Common warts are round or irregular in shape and are found most often on the fingers, elbows, knees, and face. Plantar warts have been flattened by pressure and are surrounded by hardened skin. They occur mostly on the soles of the feet and can become deeply invasive and tender. Plantar warts frequently spread within a small area of skin, forming patches of mosaic warts.

Warts are common, especially in children and adolescents, but are not easily cured. No single therapy is effective against all warts on all patients, and as a result many types of treatment are used, including the application of topical agents such as salicylic acid and essential oils as well as localized freezing (cryotherapy).

Stretch Marks

When the skin is stretched suddenly as a result of rapid growth—during pregnancy or weight gain, for example—the collagenous framework of the skin struggles to keep up, sometimes resulting in stretch marks (striae distensae). Stretch marks occur most frequently in pregnant women, when a combination of hormonal factors and increased lateral stress on the connective tissues causes pink or purplish bands to develop on the abdomen, hips, buttocks, and sometimes the breasts. Stretch marks also occur in people of both sexes as a result of weight gain. With time, stretch marks fade and become less obvious, but they never go away. Many products on the beauty market are now designed for buyers who are eager to prevent or reduce the appearance of stretch marks.

Rosacea

The skin of the face can be suffused with blood when the arteries there become dilated, or widened. When a person blushes, the face reddens and then normally returns to its resting state of blood flow after a few minutes. In some people, however, the blood vessels remain dilated for longer periods. For individuals afflicted with rosacea, the blood vessels stay permanently dilated, and the facial skin takes on the look of a red mask. The redness is sometimes accompanied by pus-filled or solid red pimples in the

FIGURE 32. The bulbous nose of the late comedian W.C. Fields was the result of rosacea, a permanent dilation of the blood vessels of the face. (Courtesy of W.C. Fields Productions Inc. and the National Rosacea Society.)

center of the face. Rosacea can be treated very effectively by gels and creams containing chemicals that reduce the inflammation, redness, and itching. Doctors can treat more serious cases with oral antibiotics. Left untreated, rosacea can progress to a condition in which the walls of the cutaneous blood vessels become distended, and the skin becomes thickened and rough. Because this situation tends to affect the blood vessels of the nose, it is called rhinophyma.

The late comedian W.C. Fields famously suffered from rosacea and rhinophyma, which led him to develop a large, thickened, and bulbous nose (figure 32). Rosacea can occur in adults of any age or skin color, but it is most common among people whose skin is lightly pigmented.

Wrinkles

Every time you move one of your limbs or part of your face, your skin moves, developing temporary creases known as dynamic wrinkles. Slowly, over the years, the skin's ability to physically rebound after being creased lessens, and some of the creases—especially on the highly mobile areas of the face—become permanent, or static, wrinkles. These wrinkles are the most evident and, for some, the most dreaded signs of aging. As the skin ages, its structural components, especially collagen and elastin, break down as the result of many natural chemical changes in the body. As elastin breaks down, mechanical stress on the skin caused by normal activity of the facial muscles causes static wrinkles to appear. Static wrinkles develop as the skin itself thins, the structural attachment between the dermis and epidermis weakens, and the elastin fibers in the skin slowly deteriorate. These events characterize the intrinsic, or natural, aging of the skin (figure 33).[13]

The development of permanent wrinkles is hastened by exposure to UVR, a now familiar culprit. Chronic UVR exposure transforms elastin fibers in the skin into broken and tangled masses of amorphous material within the dermis. In badly photoaged skin, a mattress of deranged elastin fibers crowds the skin, leading to the weathered and deeply wrinkled visage of solar elastosis. It is now widely accepted that UVR is the primary environmental factor involved in the aging of human skin and the most important external factor contributing to the formation of wrinkles. Provided they live long enough, everyone will develop wrinkles because of the inevitable aging of connective tissues, but UVR exposure accelerates damage to these tissues and greatly speeds up the formation of wrinkles. If you want to keep a relatively unlined face and neck for as long as possible, you should avoid prolonged exposure to the sun.

In the past, most cultures viewed wrinkles as signs of both advancing age and enhanced wisdom. In many industrialized countries today, however, a nearly universal emphasis on feeling and looking youthful has made wrinkles into objects of dread and derision. Youth and youthful appearance are constantly promoted by the media and corporate enterprises, as Oscar Wilde's Dorian Gray is embraced as a role model rather than reviled as a pathetic solipsist. Women, especially, are exhorted to take up the battle

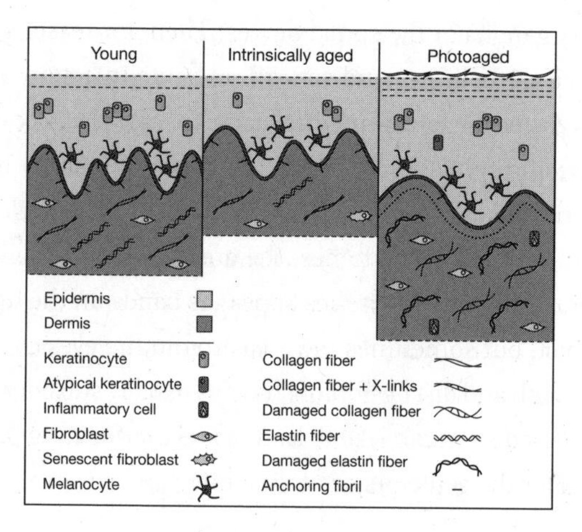

FIGURE 33. Young skin exhibits a high density of collagen and elastin fibers in the dermis and shows tight, fingerlike connections between the dermis and the epidermis (at the wavy line). Intrinsically, or naturally, aged skin has fewer such connections (anchoring fibrils), contains less collagen and elastin, and is thinner overall. Photoaged (sun-damaged) skin is thicker, is packed with defective collagen and elastin fibers, and contains numerous atypical connective and inflammatory cells, as a result of the destructive effects of UVR. The damaged elastin in photoaged skin contributes to the premature formation of wrinkles. As collagen ages intrinsically, and as a result of sun exposure, it develops more cross-links (X-links) that render it mechanically weaker and less flexible. (Illustration by Jennifer Kane.)

against wrinkles, fueling an explosive growth in treatments aimed at preventing or eliminating wrinkles and other signs of aging, a topic that the following chapter explores in greater detail.

Shingles

Viruses that cause active diseases in the body at one point can then remain dormant for years before suddenly reappearing. The varicella-zoster virus, which is responsible for chicken pox, is one example. For decades after the itchy pustules of chicken pox have disappeared, the virus can stay dormant in the sensory ganglia of the spinal nerves. Then, for reasons that are still unclear, the virus replicates and sets off an acute infection of the nerves called herpes zoster, the formal medical name for shingles. Because the virus resides in specific spinal nerves, shingles is geographically limited in its expression on the surface of the body. Outbreaks of shingles follow the boundaries of specific dermatomes, the areas of skin supplied by single spinal nerves. Most shingles rashes appear as bands on the upper torso or below the waist, but sometimes the area around the eyes can be affected. Although the rash and discomfort of shingles usually subsides within a few weeks in most people, it can lead to lasting and excruciating pain in elderly people long after the acute phase of viral infection is over.

Skin Cancers

Skin cancer is not a single entity, but rather a family of conditions caused by UVR damage to the skin's DNA. The connection between sun exposure and skin cancer was first noted among Australians of northern European (mostly English) heritage. In Australia, these people were subjected to considerably higher levels of UVR than had been the case in their ancestral homelands. Before long, skin cancers began to appear in this population with unprecedented frequency. Human skin color, with its melanin levels adapted to protect the body against particular UVR levels, evolved over a protracted period, probably thousands of years. For humans living fifty thousand years ago, journeys of fifty miles would have been rare, and those of more than a few hundred miles almost unimaginable. Nothing in our recent physical evolution prepared us for the biological challenges of living thousands of miles away from the land of our ancestors.

Skin cancers are common and are becoming more so in the aging populations of industrialized countries, especially among people whose lifestyle and travel included more and longer periods of sun exposure. As people traveled to sunny places for vacations, they were often exposed to higher levels of UVR than their bodies could handle. The "healthy tans" weren't so healthy. Now many of the people who enjoyed the healthy glow of a vacation tan are regular visitors to dermatologists. Because of the increased prevalence of skin cancers in recent years, these cancers have become the focus of intense research by dermatologists, skin biologists, epidemiologists, and geneticists.[14]

Skin cancers are classified into two major types, melanoma and nonmelanoma cancers. Melanoma is the most serious disease of the skin and is associated with a high rate of death. Nonmelanoma skin cancers—basal cell carcinoma and squamous cell carcinoma—are causes for concern but are rarely fatal.

Basal cell carcinomas (BCCs) account for 70 to 80 percent of nonmelanoma skin cancers and are the most common cancers found in humans. Most of the remaining 20 to 30 percent of nonmelanoma cancers are squamous cell carcinomas (SCCs).[15] BCCs result when changes occur in the basal layer of the epidermis (the stratum basale), which is the site of the skin's most active cell division (refer back to figure 2). SCCs are caused by changes in the squamous layer of keratinocytes in the epidermis. When damage to DNA in epidermal cells, caused mostly by UVR, is not repaired or if damaged cells are not removed from the area, cells with bad DNA are transformed and begin duplicating unchecked. At first, this duplication forms a small clone of mutant cells within the skin, but if the process goes on long enough, a tumor will become visible. These tumors usually start out small and light in color. BCCs grow more slowly than SCCs and rarely spread beyond the skin (that is, they rarely metastasize). SCCs are more aggressive and can grow beneath the skin to invade other tissues.[16]

Nonmelanoma skin cancers are rare in people with darkly pigmented skin, but they are increasing in frequency among light-skinned individuals. This trend is evident not only among older people who have had decades of sun exposure but also among people younger than forty, especially younger women.[17] This increased prevalence likely results from the popularity of sun exposure, a failure to use sunscreens and protective clothing, and a belief in the fallacy of the healthy tan, a belief held even by people whose skin is poorly adapted to withstand UVR damage.

Far rarer but more serious than nonmelanoma skin cancers are mela-

NORMAL MOLE	MELANOMA	SIGN	CHARACTERISTIC
		Asymmetry	One half of the mole does not match the other half
		Border	The edges of the mole are ragged or irregular
		Color	The color of the mole varies throughout
		Diameter	The mole's diameter is larger than a pencil's eraser

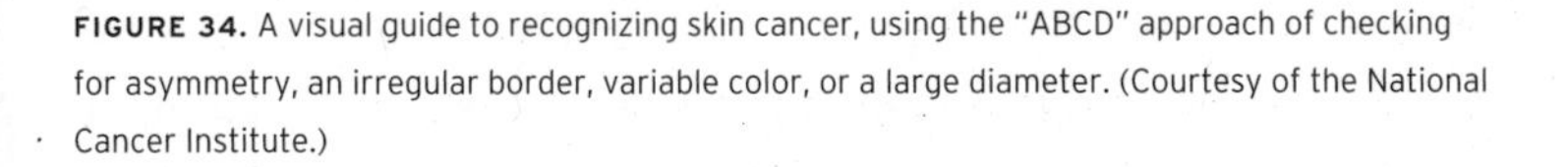

FIGURE 34. A visual guide to recognizing skin cancer, using the "ABCD" approach of checking for asymmetry, an irregular border, variable color, or a large diameter. (Courtesy of the National Cancer Institute.)

nomas themselves. In contrast to BCCs and SCCs, which affect the resident connective tissue cells of the epidermis, melanomas affect the melanocytes, which produce melanin pigment. Melanomas are much more common among people with fair skin, particularly those with very light skin, freckles, and red hair. This pattern of susceptibility is associated with variant forms of the *MC1R* gene, which helps to regulate normal pigmentation in humans.[18] Melanomas are aggressive cancers but—contrary to a widely held belief—are not always fatal. When melanomas are detected at a very early stage, before they have invaded deeper and distant tissues in the body, they are highly curable (figure 34). After they have spread to other parts of the body and established a blood supply there, they are much more difficult to defeat.[19] Melanomas often, but not always, present themselves as intensely dark lesions on the surface of the skin.

The rise in skin cancers in industrialized countries, and especially among younger groups, is of great concern because it foreshadows even greater increases in coming decades. Like other cancers that involve human behavior and the environment, skin cancer is best approached by prevention. The most effective step in prevention is avoiding excessive or unprotected sun exposure, especially if you have fair skin and naturally light-colored hair. In addition, doctors stress the importance of inspecting your own skin for precancerous changes. This is a simple process, which involves learning to recognize simple changes in the appearance of a mole or lesion on your skin. Is it asymmetrical? Does it have an irregular border? Does it show a mixture of colors? Is it large? Figure 34 provides a simple visual guide to spotting suspicious skin moles, based on images made freely available to the public by the National Cancer Institute.[20]

10
statements

As we stand naked before a mirror, our skin tells volumes about us: how old we are, the kind of life we've led, our general health, the environment our ancestors experienced. But our skin conveys much more than just the bare biological facts of our lives. Because of our unique human ability to deliberately alter its appearance, our skin proclaims our identity and individuality as we wish them to be known. For millennia, skin has served as a statement affirming an affinity to a group or a belief, as a shorthand message of how we view the world and how we wish to be viewed, even after death.[1]

Primates are visually oriented mammals, strongly so. They gather information and make decisions about animals around them mostly by observing visual cues. Does an animal belong to the same species, judging by its general appearance and pattern of coat color? Does it have friendly intentions, based on its posture and facial expressions? Does it show signs of being a potential sexual partner? As primates, we have learned to "judge a book by its cover." The ability to make rapid visual assessments of the appearance and intentions of other animals and our immediate environment is essential for survival.

In the course of our evolution, we have added substantially to this primate checklist. We also use our tactile and auditory senses as well as our highly evolved verbal communication skills to gather information about each other and our surroundings, though none of these have supplanted vision as the dominant mode of data collection. Today, humans are not just visually oriented; we are visually obsessed. In the modern realm, with the growth of digital communications media, where people increasingly learn about the world and each other through fast-paced imagery and auditory signals, appearance has come to assume an overwhelming primacy. The first impression that we read from a person's appearance—as conveyed primarily by skin (and its adornment), clothing, and jewelry—carries inordinate weight; it contextualizes and guides our subsequent interactions, often unconsciously.[2]

People use their skin as a canvas, to advertise their identity, their social status, and their social and sexual desirability. Humans have been deliberately altering the appearance of their skin for tens of thousands of years, possibly longer. Because skin is only rarely preserved in the fossil record, we can't precisely reconstruct the most ancient of these cultural practices. But the archaeological record does provide abundant evidence that humans were modifying their skin in various ways as far back as the later Paleolithic, and direct evidence from preserved and mummified skin attests to the pervasiveness of skin decoration since Neolithic times, from roughly ten thousand years ago to the present.

Body art, then, in its broadest sense has a long, complex, and interesting history. (I use the word "art" intentionally because the choices people make about decorating their skin are deliberate ones, based on a personal and cultural esthetic.) The first modifications were probably temporary markings placed on the surface of the skin, which constituted the earliest forms of painted body art and cosmetics. These were followed by more permanent modifications—tattoos, piercings, and scarification—as humans developed the technical ability to deliberately and relatively safely breach the barrier of the skin.

Modern times have brought dramatic growth in what might be termed movements of corporeal and cutaneous self-expression—the art of our bodies and our skin. These pursuits reflect the contemporary preoccupation with "body work," as people strive to transform and reconfigure their flesh through activities from dieting and bodybuilding to plastic surgery and gender reassignment, often involving sophisticated technologies or high levels of physical effort, will power, and pain.[3] In particular, the past century has witnessed startling technological developments that allow us to make personal statements with our skin. Some of these, such as plastic and esthetic surgery, grew out of the pressing need to reconstruct the appearance of individuals who had suffered severe disfigurement during the horrors of two world wars.[4] As in many other areas of medicine and technology, techniques developed to repair injuries or defects were quickly embraced for other purposes, to enhance appearance or function. In industrialized countries today, the means exist to change the basic canvas of the skin itself by adjusting its color and texture. Skin tanning and lightening, and a wide variety of methods for making the skin appear younger, are now available to those with sufficient time and money. As a result, humans of the twenty-first century can take advantage of a large and diverse assemblage of materials and techniques, accumulated over thousands of years, for expressing themselves through their skin.

Body Paint and Cosmetics

The earliest methods used to alter the skin's appearance involved the addition of color, using naturally occurring pigments. The discovery of paint pigments in archaeological contexts more than seventy-five thousand years old indicates that long before people covered their bodies with clothing, they adorned themselves with body paints.[5] This is not surprising. Modern humans, after all, hail from equatorial climes where clothing is mostly superfluous. In addition, painting the body requires only pigment and imagination, no other technology; whereas even the simplest stitched clothing requires cutting implements, awls, needles, and other paraphernalia.

Ethnographic reports of body art from all continents and Oceania record the widespread use of red ochre (hematite) as body paint. Additional colors came from yellow ochre (limonite); the black of pyrolusite (manganese); the white of ash, chalk, or lime; and other mineral pigments. Red ochre is commonly found in Paleolithic sites, and archaeologists have discovered crayonlike, honed ochre points thought to have been used for body decoration at the Middle Stone Age site of Blombos Cave, in South Africa.[6] By eight to nine thousand years ago, rock paintings, painted figurines, and incised images recording refined styles of body painting were common in northern Africa and western Asia.

Although various ochres and clays were probably the earliest sources of pigments for body painting, ground minerals and plant-based pigments such as kohl, blue woad, and urukú came into use over five thousand years ago. Kohl, a ground pigment that is usually antimony-based, derives from the Old Kingdom of Egypt, around 3000 BCE. Woad, the blue dye-stuff ancient Britons used to color their skin, was prepared from the powdered and fermented leaves of *Isatis tinctoria*. Many of the indigenous equatorial peoples of South America adopted urukú, a red coloring derived from the pith and seeds of the urukú fruit *(Bixa orellana)*. The "ecofriendly" cosmetics sold in industrialized countries today frequently contain urukú.[7]

Although traditional body painting is on the wane in most of the modern world, cultures located close to the equator, in Africa, south Asia, and New Guinea, retain the practice (figure 35). As early humans began to move out of the tropical latitudes where they first evolved, fitted clothing became an important way to buffer themselves against the vagaries of less hospitable environments. Predictably, body painting declined and was relegated to those parts of the body not covered by clothing: the hands, feet, and, especially, the face.

The first widespread use of cosmetics—body paint specially made for the skin of the face—appears to have been in ancient Egypt, where palettes of powdered, colored minerals and other compounds helped to decorate

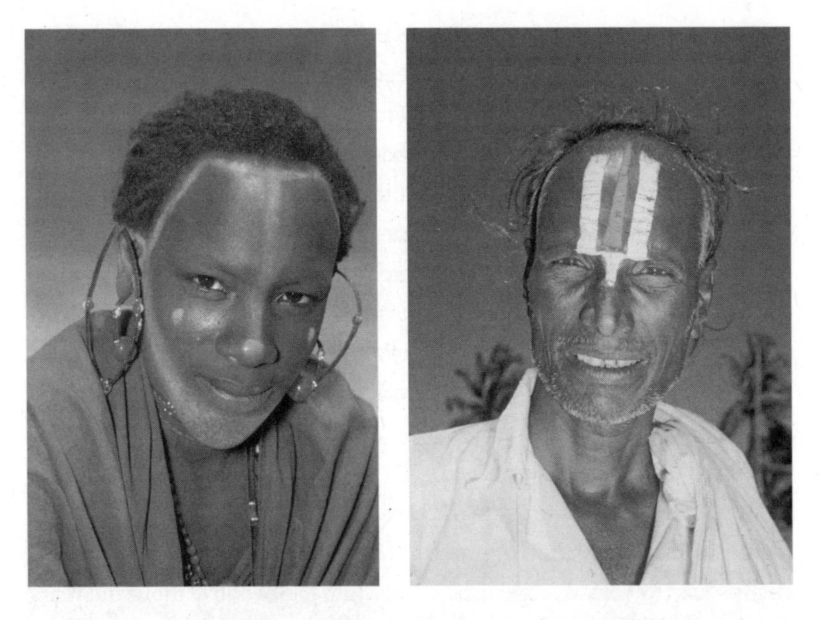

FIGURE 35. A Maasai *moran* (warrior) from Tanzania has applied a paste made of red ochre and animal fat to his face to highlight his features. A comparable pattern is seen on the forehead of a pilgrim heading to a temple in Mahabalapuram, in southern India, who has created a design of vivid stripes with white clay alongside a streak of red clay or other pigment. Similar patterns of facial decoration have developed independently in many parts of the world, used especially to highlight the forehead and the eyes. (Courtesy of Edward S. Ross.)

the faces of both the living and the dead of both sexes (figure 36).[8] The cosmetics included naturally occurring lead-based compounds as well as others that were deliberately manufactured by "wet chemistry" techniques involving controlled chemical reactions carried out in simple "laboratories." This marked the beginning of the cosmetics industry.

These early Egyptian cosmetics palettes contained kohl for darkening the margins of the eyes, malachite powder to add a green color to the eyelids, and red ochre or carmine for the lips.[9] To emphasize these highlighted features, most of the face was covered by a thick application of white lead. White lead, a mixture of lead carbonate and lead hydroxide, is familiar to us today because in years past it served as a base for household paints. In

FIGURE 36. Ancient Egyptian cosmetics, cosmetic containers, and a copper mirror are displayed here. Most Egyptian cosmetics were made from lead compounds, some naturally occurring and some specially produced. Both men and women wore facial cosmetics during life as well as after death. (Left to right: shell and mineral pigment [6-2320], a mirror [6-10236], a monkey kohl pot [6-6419], and a kohl tube and applicator [6-15175a and b]. Courtesy of the Phoebe Apperson Hearst Museum of Anthropology and the Regents of the University of California, photographed by Nina G. Jablonski.)

fact, white lead is extremely toxic, but this danger was not appreciated during the centuries when it was widely used as a cosmetic. The practice of covering the face with white lead was common not only in ancient Egypt but also in classical Greece and throughout Europe until the early nineteenth century.[10] The startlingly white faces of Japanese geisha were produced by generous applications of rice flour powder or white lead powder, which were mixed with water and made into thin pastes that were then painted onto the face. In addition to imparting a pallor, which dramatized the effect of the strong colors applied to the eyes and lips, white lead also covered wrinkles and hid unsightly scars caused by smallpox. The application of white

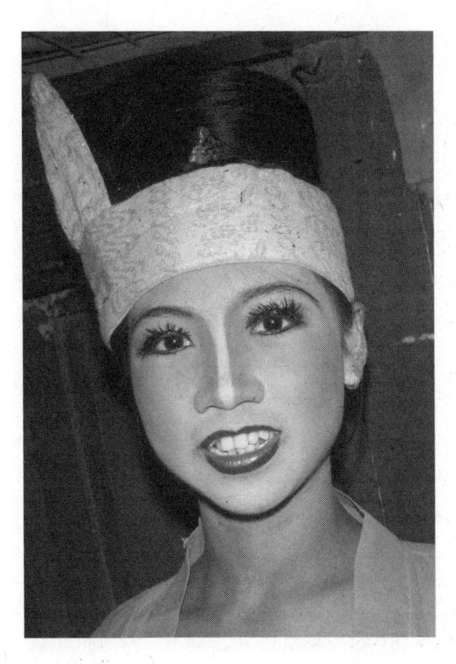

FIGURE 37. The whitened face of a female dancer from Mandalay, Myanmar, accentuates her darkened eyes and reddened lips. This use of facial makeup continues to be part of theatrical traditions in many areas of the world. (Courtesy of Edward S. Ross.)

foundation makeup to accentuate kohl-darkened eyes and reddened lips remains a staple of theatrical traditions throughout the world (figure 37).

The recent history of body painting and cosmetics in industrialized countries is complex and fascinating. In the past 150 years, a true industry has developed around the human penchant for temporary self-decoration. This industry has been fueled by, among other factors, the ability of people to see images of others (as transmitted faithfully by painting, drawings, photography, television, movies, and the Internet), the desire of observers to imitate "looks" they find appealing, and the rise of technologies capable of producing ever more sophisticated and ever changing products for effecting such looks. In recent decades, cosmetics have come to be linked with fashions in apparel, with the result that changes in style from season to season can involve complete overhauls of the appearance of the skin and all its cultural coverings. The vast majority of cosmetics products are designed and marketed for women and are geared primarily toward enhancing parts of the body that are important in conveying emotional information and sexual attraction—the eyes, eyebrows, lips, and cheeks.

Products that emphasize the size and definition of the eyes and eyebrows, beginning with kohl in ancient Egypt and extending through all manner of mascaras, eyeliners, and eye shadows in more recent times, continue to be important to females. The appeal of these products is that they accentuate the size of the eye; the distance between the eye and the eyebrow; the upturning of the outer corners of the eyes, as in a true smile; and the conspicuousness of the "eyebrow flash," the microexpression created by a brief raising of the eyebrows that signifies recognition and attention. The first two of these characteristics involve facial qualities that are used to distinguish female faces from male faces. Large eyes in adult human females are considered childlike, or neotenous, features. Children's eyes seem relatively large because of the accelerated growth of the braincase and upper face during early childhood. Among females, growth of the lower part of the face during adolescence is generally less pronounced than it is for males, leaving women with faces that are more strongly dominated by their eyes.[11]

No less valued have been products that emphasize the size and red color of the lips.[12] The lips are singularly important in verbal communication and kissing, which establish social bonds and physical intimacy with others. Cosmetics that heighten color in the cheeks are also extremely popular; radiant cheeks are associated with emotional and physical well-being, suggesting the rosiness of youth, the flush of sexual excitement, the rush of athletic exertion, or the bronzed visage of the relaxed vacationer. The earliest mass-produced rouges date from late eighteenth-century France, where they sold at an estimated rate of more than two million a year.[13] With only minor interruptions since, cosmetics use and sales have climbed steadily.

Tattoos

Tattooing is another ancient form of body art and seems to have been a near-universal human practice. Dating back at least to Neolithic times, tattooing may represent the earliest form of irreversible body decoration. The

FIGURE 38. Complex patterns of geometric designs were used to decorate both household items and people of the ancient Lapita culture of the Pacific Islands. Similar motifs are found on ancient pottery, on modern tapa cloth (barkcloth), and in tattoos. The discovery of tattoo needles at many Lapita sites attests to the antiquity of the practice among the early peoples of the Pacific Islands. (Courtesy of Patrick V. Kirch.)

oldest known tattoos are those of Ötzi, the late Neolithic Iceman described in chapter 2, whose frozen body was found in an alpine glacier. His preserved skin, nearly 5,000 years old, bore fourteen sets of permanent marks believed to be tattoos—mostly short, parallel black lines found on his ankles and back, which seem to have been produced by rubbing soot onto the skin and then puncturing the skin and pushing the dark residue into the holes (refer back to color plate 2).[14] Lavish tattoos featuring the figures of real and mythical animals decorate bodies nearly 4,500 years old that were recovered from the frozen tombs of the Pazyryk people of the Altai in Siberian Russia.[15]

The mummified bodies of high-ranking female Egyptians dating from the Middle Kingdom also bear tattoos, while a biblical injunction against tattooing dates to about this same time (Leviticus 19:28). Tattooing was widespread by three to four thousand years ago in Scandinavia, the circumpolar regions, the Americas, and Oceania; some of the most elaborate tattoos—those from Austronesia—also seem to have ancient origins. Archaeologists have recovered tattooing implements (needles and combs) from sites at the ancient Lapita cultural complex in the Pacific Islands, dating from 1200–1100 BCE, and considerable evidence from this location links the designs and motifs of tattoos with those used in the production of tapa (decorated barkcloth) and Lapita pottery (figure 38).[16]

Writings on the meanings of tattoos abound, but most authorities agree that the appeal of tattoos is that they represent a lasting inscription, conveying the importance of belonging, commemoration, and protection. Tattoos can declare a person's affiliation to a social unit. In some traditional Asian societies, for example, facial tattoos for women are considered beautiful within their own cultures but are reviled by outsiders; some groups use these facial markings as a way to bind women more closely to the group (see color plate 13). Tattoos substitute a decorated surface for the actual surface of the skin and, in doing so, transform the surface into an ambivalent statement of self-injury and self-defense. Because these decorations are usually permanent, tattooed skin can never regain the unmarked clarity of infancy.[17]

Tattooing was an integral part of most human cultures for thousands of years, but it clearly fell out of favor in most of Europe in the early Christian era, probably for reasons related to the biblical injunction against it. In subsequent centuries, the Western world came to associate tattooing either with the somewhat disreputable and marginal elements of society, such as prisoners and prostitutes, or with the primitive and the exotic, given the persistence of tattoos in other cultures around the world. Tattooing was, and in some circles still is, frowned upon in the Jewish tradition because it is considered an unholy practice.

Recent books on tattooing emphasize its ubiquity throughout history in an attempt to universalize and normalize it and render it once again attractive to modern, mainstream audiences.[18] We know relatively little, however, about the history of methods and motifs of tattooing in the West because the subcultures and elements of society with which it was associated were not well documented until roughly the last century.

Today, tattoos are the most popular form of permanent skin art, with an estimated eighty million people in industrialized countries sporting some form of "ink." The appeal of tattooing is not only its permanence but also the richness of expression that the medium makes possible (figure 39). The large canvas of the skin permits images of widely varying sizes, colors, and complexity. To the surprise of the un-inked, people who choose to get tat-

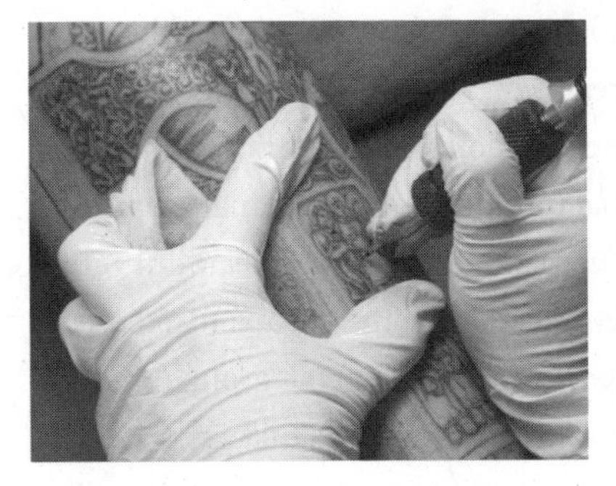

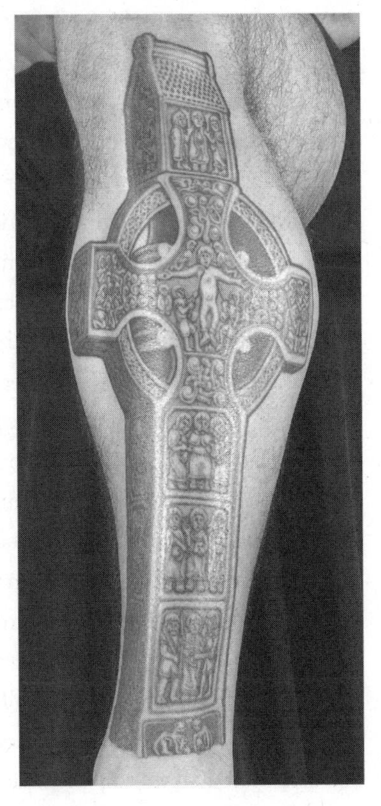

FIGURE 39. Tattooing involves the injection of small amounts of permanent ink deep into the dermis, below the surface of the skin, using a sharp, hollow needle. On the left, artist Kira Od works on the details of a complex Celtic cross design on the front of Daniel McCune's lower leg. She works under sterile conditions in order to reduce the risk of infection. On the right, the finished tattoo of the Muiredach Cross provides a good example of modern ink—rich in detail, tailored to the body, and celebrating a life-changing event in a person's history. In this case, the cross recalls a memorable trip to Ireland Daniel took with his father. (Courtesy of Kira Od.)

toos generally do not regret their decision. Most tattoos are undertaken after much deliberation and forethought, and their permanence is their strong point. Tattoo designs—once imprinted on the skin—can be indelible reminders of a significant life event. They explicitly lack the transience of the souvenir t-shirt or a temporary hair color. In an increasingly globalized world of look-alike clothing, cosmetics, and hair styles, tattoos are permanent reflections of personality, carefully calculated representations of core beliefs and sentiments that can make a uniquely powerful statement of individuality.[19]

For many, tattoos signify a permanent and visible commitment to a group or class and thus serve as a badge of affiliation or disassociation. Although they are usually acquired voluntarily, tattoos are sometimes inflicted involuntarily to mark an individual's membership in a particular group. Of all such markings, gang or prison tattoos make some of the most assertive statements of group membership. Tattoos are also often statements of romantic affiliation, incorporating a meaningful symbol or a lover's name. Individuals whose life circumstances change and who no longer wish to make such statements—they break with a gang, for example, or their romance ends—sometimes attempt to have their tattoos removed. Tattoo removal using various types of laser surgery is now possible, but the procedure is costly and lengthy.[20] When it was first introduced in the 1980s, laser removal of tattoos caused scarring in some people. New methods developed in the 1990s have eliminated this problem, but the procedure can still involve some transient loss of normal skin pigmentation in the treated area.

Modern tattoos are classified by style, with traditional, tribal, and gangster styles among the most popular. The increased appeal of tattoos since the 1980s in industrialized countries can be attributed in part to the rise of "celebrity tattoos." When cultural icons from the entertainment industry, such as Angelina Jolie and Brad Pitt, started showing off their tattoos at high-profile social events, it suddenly became acceptable, even desirable, to "have ink." As the practice of tattooing spread, a specialized vocabulary developed to describe individual designs and styles: now terms like "darkside," "old school," "jailhouse," and "wild style" evoke specific styles of imagery. Many of the people who get tattoos in affluent countries today exercise great care in choosing the design and composition of their tattoo and the artist who will create it, sometimes going to great expense and traveling long distances to engage a famous tattoo artist in another city or to attend one of the many newly popular body art fairs.

The use of skin transfers ("temporary tattoos") or the painting of semipermanent pigments on the skin can achieve esthetic effects similar to those of tattoos. The traditional art of henna skin painting, or *mehndi,* belongs to the latter category. Mehndi began as a traditional means of decorating a woman's hands and feet with complex designs painted with henna dye. The

FIGURE 40. Mehndi artist Ravie Kattaura paints the skin with henna, which imparts a semi-permanent design that usually lasts for seven to ten days and sometimes for as long as six weeks. In traditional contexts, the ceremony of applying the mehndi is as important as the effect of the finished design. (Photograph by Caroline Kopp, © 1992 California Academy of Sciences.)

tradition originated in North Africa and the countries of the Middle East and was brought to India in the twelfth century by the Moghuls. Originally reserved for brides-to-be, mehndi has become popular in recent decades as a mode of self-decoration among a wider spectrum of women in Asia and the West (figure 40). Mehndi typically lasts on the skin from seven to ten days and sometimes as long as six weeks, depending on how soon after application the skin was washed and how often it is washed in subsequent weeks. The application of a complex henna design, considered an important beautification ritual, can take up to six hours, especially for a bride.[21]

Piercing and Scarification

Piercing, scarification (cicatrisation), and branding are widespread practices found in indigenous cultures on all continents, suggesting an ancient origin. Piercing likely provided a built-in means of affixing decorative and pre-

FIGURE 41. Traditional body piercings include extremes of self-mortification such as piercing the skin in the chest and back with multiple hooks, as Hindu fakirs sometimes did, and suspending the body weight by hooks forced into the chest wall, as in the famous suspension ritual, or *okipa* ceremony, of the Mandan people, captured here by artist George Catlin. This 1867 painting depicts a ceremony that took place around 1835.

cious objects to the body in early cultures where clothing was not widely used or needed. Whereas tattooing is more common among people with lightly to moderately pigmented skin, scarification and branding are seen more frequently among those with darkly pigmented skin. Highly melanized skin tends to form more prominent keloid scars, whose raised pattern is quite visible on the dark skin.

Like tattooing, piercing has become more popular in industrialized countries in the past twenty years. Long considered in the West to be the preserve of women who wanted a permanent method of affixing jewelry to their ears, piercing has become literally and figuratively more widespread, involving many parts of both the male and the female body. Piercings on parts of the face other than the ears are the most popular forms; they represent an extension of the trend toward multiple ear piercings that became popular in the 1980s. Facial piercings have become so fashionable and commonplace in the past decade that simple ones are performed in kiosks at shopping malls.

Piercing other parts of the body was first associated in the West with sadomasochistic relationships and the modern primitive movement of the 1970s, which sought to overcome the numbing effects of modern technology and industrialization on the senses by encouraging permanent body modifications as forms of spiritual exploration and personal expression. These practices derived their inspiration from Asian fakirs and Native American warriors who sought to express their spiritual power through feats of self-torture, such as piercing the flesh with weighted hooks or suspending the body from hooks (figure 41). Punk-rock performers of the 1970s and 1980s emulated more traditional feats of piercing. By making self-injury part of their acts, they sported extreme body modifications in order to make "anti-fashion" statements offstage.[22] A different philosophy is embraced by the performance artist Stelarc, who has performed body suspensions as part of his continuing effort to escape the confines of the skin and eliminate the barrier between the public space and the inner space of the body (figure 42). To him, the suspension is "a manifestation of the gravitational pull, of overcoming it, or of at least resisting it. The stretched skin is a kind of gravitational landscape."[23]

Like tattooing, body piercing has given rise to its own jargon to describe types, styles, and regions of piercing. Intimate body piercings involving the nipples and genitalia of both sexes are becoming more common as individuals seek alternative forms of sexual expression as well as the expressions of personal and group identity common to other forms of permanent body art.[24]

Scarification is a decorative procedure based on the body's own penchant for building visible scars—especially the highly melanized keloid scars found among dark-skinned people—when the skin suffers a severe burn or a deep cut. Scarification can be created by branding or cicatrisation. In branding, hot pieces of metal that have been fashioned into a desired shape are heated and applied to the skin in order to create a second- or third-degree burn. Because the scar resulting from the burn is typically much larger than the original lesion, allowing for less intricacy, the designs tend to be much simpler than those used in tattoos. Some body artists have recently introduced the use of surgical lasers to inscribe "brands."[25] Cicatrisation is performed by

FIGURE 42. The work of performance artist Stelarc deliberately questions our notion of the skin as a meaningful barrier. Stelarc has performed twenty-seven suspensions with the insertion of hooks into the skin in different positions on his body. These suspensions were intended to convey the human yearning to float or fly. In the artist's words, "the cables [leading from the hooks] were lines of tension which were part of the visual design of the suspended body." He described his stretched skin as "a kind of gravitational landscape." (Spin Suspension, Artspace, Nishinomiya, 1987, © Stelarc, photo by H. Steinhausser. 1995 interview with Paolo Atzori and Kirk Woolford, www.ctheory.net/articles.aspx?id=71.)

specially trained experts, who cut the skin with a sharp knife or a scalpel, without introducing ink. Highly intricate designs can be created when small cuts are incised in the skin using razors or thorns and the wounds are then covered with charcoal to produce a pattern of raised scar tissue (figure 43).[26] The charcoal (or other irritant) introduced into the cut creates a permanent scar that never fades, although it becomes slightly less obvious over time.

Decorative scars are used to document and record important stages in a person's life. Such ritual scarification was and still is a culturally sanctioned practice associated with the transition to adulthood in many indigenous cultures around the world. Although traditional scarification is waning in popularity as Westernized standards of appearance begin to encroach on non-

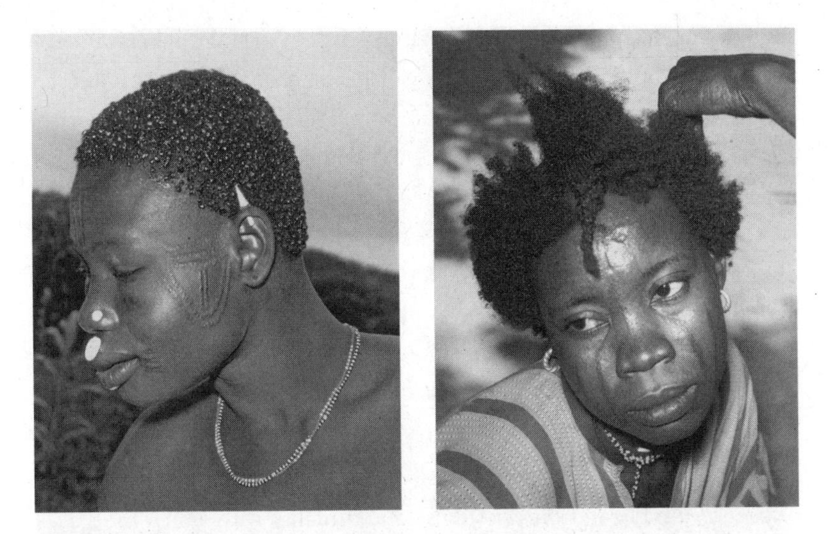

FIGURE 43. Delicate, decorative scars adorn the face of a Nubian woman from the Mandara Mountains on the border of Nigeria and Cameroon (left). Such scars are produced by cicatrisation—scoring the skin with a sharp knife and then introducing charcoal or another substance into the cut to create an obvious scar. The deep vertical scars on the cheeks of a Kanuri woman from Damatura, Nigeria (right), were made by clean and even cuts into the dermis. In both cases, scars were made on the young women's faces to herald their entry into adulthood and to accentuate their beauty. (Courtesy of Edward S. Ross.)

Western cultures, it is still practiced in parts of Africa and on some Polynesian islands. The pain associated with the act of incising the skin is considered an important part of the ritual because it creates an indelible memory that scores the consciousness much as the knife cuts the skin, emphasizing that the individual's childhood is now in the past.

Beyond Pale

For centuries, judgments regarding human character, potential, and desirability in different populations have been based on skin color. In the centuries during which European powers colonized Africa and used the continent as a source of raw materials and labor, dark skin was associated with a suite of undesirable personality traits and moral deficiencies in a self-serving effort to rationalize the trade in human slaves. From the Enlightenment until the nineteenth century, European and American philosophers and natural historians espoused ideas about the genesis of skin colors and then generalized their conclusions to entire populations. In Lorenz Oken's early nineteenth-century *Handbook of Natural Philosophy,* he distinguished between what he called the Moorish ape man and the human white man. The former was black and unable to display his inner emotions because of the obscuring influence of color. The latter was white and transparent, with skin that allowed his interior emotions to shine through. The absence of visible blushing in darkly pigmented people was seen as proof that they were devoid of moral feelings, especially shame. Dark-skinned slaves were thus stripped of the sense of morality that defined humanity and relegated to subhuman status. This perverse logic was used to justify the continuation of the slave trade and the harshness and brutality with which the slaves were treated.[27]

The social desirability of light skin in European cultures is a long-standing legacy. A preference for lighter-skinned individuals appears to have existed within some populations of generally dark-skinned people in Africa and Melanesia well before European contact, although the opposite is the case in some of these societies. Positive social links between lighter complexions and enhanced status and potential are referred to as "colorism," or stratification by skin color. The connection of light skin with infancy and femaleness (discussed in chapters 5 and 6) may have been a precursor of the cultural notions that associate light skin with the innocence of youth, heightened femininity, and cosseted privilege. It also became socially desirable to spend more time indoors at leisure, distant from hard physical labor outdoors that might darken the skin. Possessing light skin, then, came to be associated with membership in a higher-status, more fashionable class and—especially among women—was viewed as an emblem of sexual desirability. The cultural construction of whiteness was, and still is, pervasive.[28]

This ideology has not only fundamentally influenced personal esthetic choices; it has also promoted an industry devoted to the temporary and permanent lightening of human skin. In many Asian countries, most women diligently avoid spending time in the sun and use skin lightening agents to further whiten their skin. In countries where constitutive pigmentation is generally darker, skin bleaching agents (mostly hydroquinone formulations, but also toxic mercury-based products) have become increasingly popular and now account for a large fraction of total cosmetics sales.[29] Fueled by global marketing, the social desirability of white or lighter skin is being promulgated throughout more and poorer countries, including those of equatorial regions, where dark pigmentation provides important protection against high levels of UVR. In multicultural countries, aggressive marketing of skin lightening products has also promoted the spread of colorism by developing ideals of lightness, a pernicious and disturbing social trend.

By the 1950s and 1960s, the majority of nonagricultural working people in the Northern Hemisphere stayed indoors most of the time, in the home, the office, or the factory. It was at this time that a contradictory social phenomenon arose: among those of European descent, tanned skin began to be seen in a more positive light, largely because female celebrities—following the early lead of fashion trendsetter Coco Chanel—sported their "holiday tans" in public. Rather than being associated with the rigors of outdoor toil and the hard labor of the poor, skin darkened by sun exposure was heralded as a "healthy tan," associated with luxury vacations in sunny resorts and the leisure of sunbathing. Suddenly, a look that had been reviled for centuries became chic because it was unusual and a sign of privilege.

The irony of present-day skin color aspirations is not lost on most people. Naturally dark people in many parts of the world are increasingly seeking ways to lighten their skin, while the naturally light-skinned are trying to find new ways to darken theirs. The paradox of these aspirations speaks to the influence of skin color on people's perceptions of themselves and one another, and to the potent influence of images associated with elevated social status.

Today, tanning remains a popular pastime, whether the tan results from toasting in the sun on a beach, lying under the UV bulbs of a sunbed in a tanning salon, or applying darkening chemicals to the skin. The impact of tans produced by natural or artificial UVR is still being felt in the enormous personal and social costs of prematurely aged skin and skin cancers, especially among lightly pigmented inhabitants of Australia, Florida, and the American Southwest. For the past twenty years, medical authorities have cautioned against tanning because of the explosion in skin cancer rates resulting from increased UVR exposure. There is also a probable link between recreational tanning (tanning under natural sun or artificial UVR), folate depletion, and birth defects that is worrying for women of reproductive age. Despite these substantial risks, many still view tanned skin as fashionable and as a sign of well-being, spurring the development of a large indoor tanning industry in the Americas and Europe.[30]

For those who want the look of tanned skin but recognize the risks of UVR exposure, simulated tanning, or self-tanning, has now become a popular option. A person can create an artificial "tan" by applying lotion containing dihydroxyacetone (DHA) to the skin. DHA reacts with components in the epidermis, producing "melanoid," or melanin-related, compounds that impart an orange-brown color to the skin a few hours after the lotion is applied.[31] The fake tan fades when the treated stratum corneum of the epidermis is shed. The desire of millions of pale-skinned people to appear tanned without risking skin cancer has now spawned a multimillion-dollar industry in simulated tanning, involving nearly all the major manufacturers of cosmetics and body lotions in Europe, the Americas, and Australia.[32]

Altering the Canvas

German novelist Franz Kafka lamented that the raiment of skin was the only outfit that humans had to wear for a lifetime. He faced with resignation the process of growing old and becoming wrinkled and viewed the skin "not only [as] a garment but also a straightjacket and fate."[33] Kafka might well have appreciated today's advances in medical and cosmetic technology that have made it possible not only to change the appearance of the skin through cosmetics but also to revivify its physical basis. With longer lifespans, maintaining a youthful look goes hand in hand with maintaining youthful levels of physical and sexual activity. Concern over young-looking skin was once the province of mostly middle-aged women, but it has now spread to men and to teenagers and young adults of both sexes.

Perusing popular magazines for women and men quickly reveals the wide array of techniques now available to recapture the skin's youthful appearance or to slow the process of aging. For many years, face lifts performed by plastic surgeons were considered the primary means of tightening aging skin (mostly on the face) and subduing wrinkles. While face lifts of many types are still considered the gold standard of cosmetic dermatological procedures, the development of new surface treatments and injectable compounds for changing the skin's appearance has revolutionized the field.

Since about 1990, skin rejuvenation has been greatly democratized. Many new skin treatments are less expensive than face lifts, allowing middle-class individuals to pursue the ideal of youthful skin, which had previously been possible only for the very wealthy. Equally important, celebrities who have "had work done" are no longer the butt of jokes but are often elevated to the status of role models by the media. Television programs like *Extreme Makeover,* which showcase sometimes drastic surgical and nonsurgical cosmetic procedures, have made appearance overhauls and skin enhancement into spectator sports. The variety of the procedures and materials that claim to revivify skin is vast and growing, and this growth is fueling expansion of lucrative branches of medicine and applied chemistry. According to the American Society for Aesthetic Surgery, Americans spent nine billion dollars on plastic surgery in 2004. The National Consumers League indicates that nearly ninety million Americans have used or continue to use products or procedures in an attempt to reduce the visible signs of aging.[34]

From the point of view of evolutionary biology, the most interesting of the newer procedures involves injecting denervating agents into the skin in order to prevent the formation of facial wrinkles and to reduce the severity of existing ones. In botox procedures, as they are commonly called,

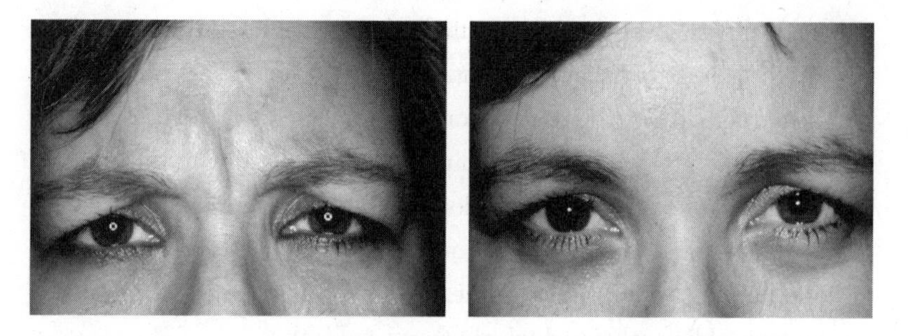

FIGURE 44. Before botox treatment (left), the woman in the photograph exhibited a pronounced crease between her eyebrows when she frowned. After treatment (right), this crease was far less noticeable when she wore the same expression, because the botox prevents the facial muscle from contracting. (Courtesy of Alastair Carruthers.)

purified botulinum type A neurotoxin is injected into various muscles that control facial expression. The toxin works by blocking the action of the neurotransmitters at the neuromuscular junction, thus preventing contraction of the muscle. The effects last only three to four months and must be repeated to prevent the muscles from resuming normal contractions. Repeated injections at the same site eventually lead to muscular atrophy and a reduced ability to produce facial expressions and the dynamic wrinkles resulting from them. Used initially for reducing the severity of furrows in the brow and "crow's feet" around the eyes (figure 44), botox is now often used on the lower face and neck to reduce the appearance of deep nasolabial folds, between the nose and the corners of the mouth, and prominent "cords" on the neck produced by the sheetlike platysma muscle that covers the front of the neck.

The use of botox represents an interesting way in which a competition plays out between compelling forces in primate evolution. On the one hand, a premium is placed on youthful appearance and its association with greater fecundity; on the other, enhanced communication through the subtleties of facial expression is of value. Taken to an extreme, botox all but eliminates facial affect, leaving the skin of the face passive and lifeless even during animated speech. This makes the cognitive interpretation of information genuinely difficult, especially when the emotions expressed on the face do not match those being conveyed by language. For example, public figures, actors, and celebrities who use botox may be called on to give impassioned speeches, which they hope will inspire action or empathy. In the absence of a normal range of facial expressions, they will inspire neither. Of particular concern in a forensic context is that botox can be misused as a tool of deception because it can alter the facial expressions that people often rely on to establish relationships of trust.

Most of us at one time or another have made some kind of personal statement with our skin, even if it was only transitory—a smear of lipstick, or perhaps a temporary tattoo. Many people are motivated to engage in daily enhancements such as using cosmetics; others seek permanent or semi-permanent alterations of color or texture through tattooing or cosmetic surgery. Still others seek to make fundamentally different statements with their skin. The realm of the performance artist represents an alien world to many, but it is noteworthy because of the ways in which some of these artists challenge widely held beliefs about culturally conditioned concepts of corporeality and individuality. The artist Orlan uses her own face and body as a canvas, with plastic surgery as her medium of choice. She aspires to become a "synthesis of classical beauty" by undertaking surgeries that will reconstruct in her the traditional esthetics of female beauty. In what she refers to as her "Carnal Art," her objective is not to become beautiful but rather to suggest that the objective of beauty "is unattainable and the process horrifying."[35] By means of cosmetic surgery, Orlan has altered portions of her face to simulate the features of several historical icons of feminine beauty created by men, including the forehead painted by Leonardo Da Vinci in the *Mona Lisa* and the chin of Botticelli's Venus. To paraphrase Orlan, the body and the skin exist to be altered at will, and it is outmoded to accept one's natural appearance.

11
future skin

Much of what this book conveys can be distilled to two essential observations. The first is that human skin shares most of its biological properties with the skin of our primate and mammalian relatives. It differs little from that of monkeys and apes, and the ways in which it is different—its sweating ability and its coloration—have evolved since we last shared a common ancestor with chimpanzees, some six million years ago. The second observation is that human skin is unique because of what people do with it. Among other animals, the skin and its appendages, including scales, feathers, and fur, serve to advertise an animal's anatomy or prowess. But the human penchant for decorating the skin is unmatched in the animal world. A peacock can flaunt its feathers, but it can't change them for every show.

The colors or pictures we apply to our skin communicate our values and aspirations as well as our hopes and personal histories. Even when we adopt the "natural look" and don't adorn our skin at all, we are making a social statement. Our skin talks even when we don't; it is not a neutral canvas. Through the expressive functions of skin and body decoration, we have expanded the communicative potential of our bodies and reinforced the primacy of the visual sense in our sensory repertoire. Especially in industrialized societies, this may well be a response to the increasing importance of the sense of self and the identification of self at the level of the skin. It also reflects the heightened human capacity to recognize self in others and to interpret and internalize the complexly coded visual signals that others present to us through their appearance.

As the most self-involved and manipulative of primates, we will continue to change ourselves to an even greater extent than we change the world around us. When it comes to our skin—our protective envelope, billboard, and largest sensory portal—our scientific and artistic creativity knows few bounds. We cannot predict our future skin in detail, but major trends in medicine and art provide us with some leads. The functions and potential of skin will be expanded along at least three major frontiers in the next several decades. Advances at each frontier will be made by very different constituencies, some strongly corporate, others highly individualistic and anarchic, sometimes involving new combinations of scientific, technological, and artistic expertise.

The first frontier, and the easiest to envision, is alteration of the biological functions of skin in order to treat diseases and injuries. Like many innovations in medicine, these advances will be introduced initially as cures for specific problems—the treatment of burns, for example—and then developed into enhancements of normal function. Progress in this area will largely be driven by the work of specialists in gene therapy, molecular biology, pharmacology, and bioengineering.

The second major frontier will focus on communication through and with the skin. Here, some innovations, such as implanted sensors and communication devices, will be primarily practical; others, such as new types of cosmetics and skin colorings, will be mostly esthetic. New combinations of practical and esthetic considerations are likely to yield the most surprising changes in human skin-based communication, changes that will be startling in their rapidity and effect. Some of these advances will involve projecting visual imagery on the skin or implanting in the skin sensors and devices capable of communication and stimulation. Ever more sophisticated and interactive virtual reality experiences will become part of our everyday routines, and entertainment based on the interaction of implanted devices with visual and auditory devices in the environment will become commonplace.

The third major frontier for future skin will be the creation of "skin" for robots that will allow them to simulate human touch. This work will be carried out primarily by computer engineers and psychologists, far removed from the conventional theaters of dermatology and natural science. These and other innovations will extend our current thinking about what skin is and what it is capable of and will challenge our very notion of corporeality.

Healing

Serious traumatic injuries to the skin, such as full-thickness burns and multiple lacerations, are less common today than they were in the past. They still occur in heartbreaking numbers, however, and leave physical and emotional scars that can be disabling in every sense of the word. It is no wonder, then, that the past ten years have seen an intensified focus on research designed to improve the treatments for serious burns and scars. Although our skin's ability to heal after an everyday cut or abrasion is remarkably good, human skin has a relatively poor capacity to heal itself after extensive burns. This difference likely developed because burns were not common injuries in our evolutionary past. Animals constantly sustain abrasions and lacerations in the course of interacting with the environment or with one another, but burns are relatively rare.

One of the major problems involved in treating large burns and wounds is reestablishing the natural laminar structure of the skin in order to regain its mobility and elasticity. The scar tissue that replaces skin in large burned areas can develop into a massive, inelastic, and confining mat composed mostly of a dense and ever-tightening meshwork of collagen. When scar tissue spans large areas or crosses normally mobile joints, it can become a grotesque and immobilizing straightjacket. Today, treatment of deep or extensive burns often uses skin grafts, in which thin strips of a person's healthy skin are harvested from a donor site, taking mostly epidermis with a little dermis. These strips are then stretched over the exposed area, while the donor site is left to regrow naturally.

In the past decade, medicine has come up with an alternative to skin grafts. This new technique uses sheets of cells cultured from a patient's own skin. The main problem with this technique is time. Sheets of cultured skin cells take three weeks to grow, a period during which life-threatening infections can endanger the patient. Fortunately, new methods for speed-culturing skin cells and spraying them onto wound surfaces are being researched, to be used either independently or in conjunction with skin grafts.[1] Such procedures at least seal the wound against the environment and significantly reduce the chance of infection, although they still result in scarring.

Research is now going forward aimed at reconstructing the dermis and the epidermis in a single step. This can be done by spraying cultured skin cells (keratinocytes) onto artificial "scaffolding" made of animal cartilage and collagen and then applying the composite sheet to a person's wounds. The dermal cells move into the scaffolding to rebuild the dermis, while the keratinocytes migrate to the surface to form the epidermis.[2] Although considerable testing and refinement remain to be done, this technique—in theory at least—holds great promise.

These "bricks-and-mortar" approaches have been highly successful in helping to reclaim large areas of skin lost through burns and major surgeries, but the search for natural and cost-effective skin substitutes is also taking a new direction. Since the introduction of cell-based skin substitutes, research has focused on making such products more sophisticated. Imagine, for instance, applying skin cells composed of not only a burn patient's own epidermal keratinocytes but other skin cell types as well, so that more natural skin could more quickly regenerate. This method could reduce or eliminate the need to use skin grafts or artificial dermal scaffolding derived from animals, and it could also help to solve the problems that arise when a patient's immune system tries to reject foreign grafts. In coming years, gene therapy will be further combined with tissue culture methods to develop fast-growing and naturally laminated substitutes for skin.[3]

Knowledge of the genetic basis for skin diseases will become increasingly important in treating these diseases. Psoriasis is an excellent case in point. It is a common disease characterized by localized inflammation of the skin that results in red, scaly, and sometimes disfiguring lesions on the skin's surface. In some people, psoriasis also expresses itself as arthritis in the small joints of the hands and feet.

In individuals who suffer from psoriasis, a particular class of proteins, known as *Jun* proteins, is in short supply because of altered function at the *PSORS4* genetic locus. This deficiency appears to trigger a cascade of biochemical events in the skin, increasing production of inflammatory compounds and recruiting specific types of white blood cells to specific areas.[4] These events eventually cause the changes in the skin's surface appearance and the itchy discomfort that are part of psoriasis. In other words, genetically based changes in the composition of the epidermis initiate the disease. Now that the genes and proteins responsible for psoriasis have been tentatively identified, it should soon be possible to use drug therapy to block the expression of the defective skin genes or to produce enough *Jun* proteins to arrest the inflammatory cascade.

Even the treatment of scars will eventually benefit from new genetic information, specifically data gained from studies of the genetic control of embryonic development. When an embryo is deliberately wounded in an experiment, it is able to heal without a scar. This makes sense, because embryos must constantly close holes in themselves during normal growth. The challenge becomes determining how embryos do this and how we can selectively reactivate this ability to assist in scar-free wound healing after birth. Considerable attention is now focused on a process in embryonic skin called the JNK cascade, which plays an important role in the migration of skin cells in embryos during normal development and after wounding.[5] The genes switched on by this cascade appear to be activated by mechanical stress on cells. Identifying the genes underlying the JNK cascade should help researchers discover new therapies for skin healing, which are desperately needed for chronic wound healing (treating leg ulcers in people with diabetes, for example) and burn treatment. Thus, we can look ahead to a time when, at least for some, scars will become a thing of the past.

Communication

Advances in treatments for damaged or diseased skin will be dramatic, but they will be overshadowed in the public arena by new ways of enhancing the skin's appearance and making it into an even more dynamic surface for the communication of experience and aspiration. Trends that emphasize the value of youthful skin, the use of cosmetics and cosmetic surgery, custom skin coloring, and various modes of skin decoration show no signs of abating, as chapter 10 points out.

It is easy to imagine, for instance, that gene therapy may someday be able to prevent or reverse aging processes in the skin. Less in the realm of fantasy are new injectables, especially for the face and neck, that are designed to stimulate epidermal regeneration, tighten up sagging folds, or plump up incipient creases. Cosmetics will undoubtedly become increasingly chemically sophisticated, with even greater emphasis on color enhancement delivered along with therapeutic agents ("cosmeceuticals") or with new compounds designed to change the mode of light reflectance from the surface of the skin.

Methods of changing skin color, whether permanently or semi-permanently, will become more varied and sophisticated. Ever more natural tans will be created by tanning and sunscreen products containing melanin or melaninlike pigments, as well as by activation of the body's own melanin production in the absence of sun. Or, in contrast, people will be able to bleach their skin by reversing the same process and deactivating melanin production at the cellular level.

Tattoos will become even more popular when less expensive and laborious methods of tattoo removal become available. We can envision tattoo inks that can break down into harmless and invisible by-products after a defined period of time or after sustained exposure to a single wavelength of visible light, thus obviating the need for painstaking laser removal of embarrassing names or incriminating symbols.

Our consciousness of physical appearance and modern society's emphasis on our skin as a billboard for self-advertisement will only continue to grow. The billions of dollars currently spent each year on enhancing the appearance of skin may seem paltry in just another decade or so, as we strive to make ourselves appear more youthful, interesting, or singular by modifying our skin.

The coming years will also witness increased emphasis on enhancing and expanding the skin's communication functions by computerized information transfer using direct and remote touch. The deliciously acute sense of touch that we share with other primates will be rediscovered and stimulated in new and unforeseen ways. Already we can implant information-bearing microchips (radio frequency identification, or RFID, chips) in the skin, a technique that is being widely used to identify pets, livestock, and human medical patients.[6] The idea of implanting microchips carrying personal information has been slower to catch on among the general populace for a variety of good reasons, mostly related to basic civil rights.[7] But it is easy to imagine that people will soon be wearing a variety of such devices, some implanted voluntarily and temporarily, some not. It is already possible, for instance, to pay for an evening of partying at a nightclub in Spain by drawing against your prepaid account—an account whose details are held in a tiny microchip inserted under your skin at the beginning of the evening, so that you needn't carry a credit card or wallet.[8] Many other types of personal information could be contained in embedded microchips, everything from your own DNA sequence to your entire medical or credit history or a record of criminal convictions.

Perhaps less worrisome would be devices able to transmit or receive information concerning conditions within your body or the environment. Implanted temperature sensors could, for example, transmit data about your surface body temperature to "smart clothing" that would warm or cool you where needed. You could be automatically alerted to high levels of ultraviolet or ionizing radiation or to the presence of strong magnetic fields, thus preventing damage to the skin or malfunction of medical devices such as pacemakers and insulin pumps.

One of the most exciting uses for embedded microchips may be in the realm of entertainment and performance. Imagine performers with implanted microchips controlling lighting effects, sound levels, projected imagery, or even aroma generators by transmitting automatically detected or preprogrammed information to lighting, sound, or special-effects systems in theaters. The possibilities for benign, malign, and artistic applications are nearly endless and undoubtedly will be explored vigorously in coming decades as ever-smaller and more biocompatible chips are developed.

Pressure, temperature, and conductivity sensors implanted under the skin and in devices next to the skin will also radically change the personal sphere of touch and touching. Instead of being restricted to touching and responding to the touch of someone in close proximity, you might be able to engage in remote touch and even generate a remote physical presence. Today's virtual reality simulations are somewhat crude and cumbersome, but they do allow people to experience a wide range of real-world situations by delivering a combination of visual and haptic (touch-based) stimuli in highly controlled environments. The ability to generate presence through multisensory virtual reality devices is proving useful today in many therapeutic contexts, such as the treatment of phobias, social anxiety, post-traumatic stress, and pain management.[9]

Advances in computer and artificial intelligence technology will soon make it possible for us to build highly sophisticated interactive systems that can be harnessed in many relatively benign contexts, including medical care, psychotherapy, physical intimacy, and general entertainment—think, for example, of providing soothing touch to a group of invalid elderly individuals, offering a long-distance hug, enjoying a remote interactive grope with your latest chat-room hookup, or engaging in a bit of harmless virtual bondage. These scenarios are closer to becoming reality than you may think, as the capacity to deliver a "cyber caress" to an animal now exists, paving the way for a "hug suit" for humans that can deliver a soothing hug to another person via the Internet.[10] The same technology will, of course, also open the door to more sinister applications, such as espionage, abusive interrogation, virtual molestation, and torture. The possibilities are both captivating and frightening and are likely to challenge our basic notions of self, physical presence, and personal responsibility.

E-Skin

Another area of skin-related research in which we will see rapid developments in the next decade is electronic skin. When we do routine things like cracking an egg, turning a doorknob, shaking hands, walking barefoot on a carpet, or squeezing a tube of toothpaste, an array of pressure and temperature sensors in our skin and positional sensors in our joints tell us just how hard we should squeeze, how we should tread, and where our limbs should go. When robotics experts turned their attention to designing robotic bodies that could operate in the kinds of environments in which people live and perform real-world actions like opening a bottle, lifting a person into bed, or answering a door, they ran into serious problems. How does a robot feel what it's doing so that it doesn't break the bottle, hurl the person over the bed, or wrench off the doorknob? The exquisite sensitivity, natural stretchiness, and pliability of human skin are tricky to replicate. Advances in the development of artificial electronic skin for use in robotics have been slow in coming because the complex and economical design of real skin allows it to do its many jobs with only a few millimeters of thickness.

But the future of social robotics lies in solving the engineering challenge of human touch through the invention of accurately replicated, humanlike artificial skin. Several types of artificial electronic skin are in the development stage, some with a "natural" degree of stretchiness, some lacking this characteristic (see color plate 14).[11] But the hope is to impart to robots some of the feeling—literally—of being human so that they can more effectively imitate human activities and interact with people. The invention of good artificial skin may also herald something quite unexpected: the birth of a sense of self. If a robot is covered with skin that allows it to distinguish between its own surface and that of something in its environment, it is developing a consciousness of the boundary between itself and the environment, of an inside and an outside.[12] The making of truly humanlike skin will bring us much closer to the development of truly humanlike robotic beings.

OUR FUTURE SKIN will continue to do what skin has been doing for us for millions of years, but it will also do much more, as we use technology to extend our touch and consciousness literally beyond our own skin. We as humans enjoy our skin probably more than we enjoy any other part of our bodies. We benefit greatly from the physical stimulation we receive directly from the skin, we derive so much information from our skin, and we take such pleasure in beholding its visual and tactile charms. We fear it for the same reasons—because it is the interface to physical and psychological intimacy. The skin mirrors our self-image and social importance. Through the skin, we project an image of ourselves to others and announce our relationship to particular groups. Consciously or not, most of us update that image regularly as we seek to be identified with or distanced from various groups. Through our naked, sweaty, marked-up skin, we tell the world who we are. Our skin is us.

GLOSSARY

ALBINO: An individual characterized by a lack of pigment in the skin, hair, and/or eyes as the result of a disruption in the synthesis of melanin. The main forms of albinism are ocular (affecting only the eyes) and oculocutaneous (affecting the eyes, skin, and hair).

AUSTRALOPITHECINE: Referring to the ancient hominid genus *Australopithecus* or to a grade of early hominids preceding the genus *Homo*. The so-called robust australopithecines are distinguished by having larger teeth and a more heavily built jaw structure than the species known as gracile australopithecines.

BIPEDALISM: The ability to stand or walk on two legs rather than four (quadrupedalism). Humans exhibit habitual bipedalism—that is, they stand and walk on two legs all the time. Habitual bipedalism evolved independently in several lineages of vertebrates, including the ancestors of certain lizards and dinosaurs and the ancestors of birds and humans.

CATARRHINE: Referring to the Catarrhini, the infraorder of primates containing the Old World monkeys, apes, and humans.

CICATRISATION: Incising the skin in order to produce decorative scarring, sometimes involving the introduction of an irritant such as charcoal into the wound to create a more pronounced raised scar.

CONSTITUTIVE PIGMENTATION: The skin coloration determined by a person's genes, best observed on an area of skin not often exposed to sunlight, such as the inner side of the upper arm.

CUTANEOUS: Of or pertaining to the skin.

DERMIS: The deeper of the two major layers of the skin, composed of a dense network of collagen fibrils and elastin fibers, interspersed connective tissue, immune cells, blood vessels, and nerve endings. The dermis imparts toughness to the skin.

ELASTOSIS: The breakdown of elastin fibers in the skin. When this breakdown is caused by UVR exposure, it is known as solar elastosis.

EPIDERMIS: The outermost layer of the skin.

EPITHELIUM: The general term used for the covering of internal and external surfaces of the body, including the lining of the mouth, blood vessels, gut, and other organs.

EUMELANIN: The most abundant type of melanin in the human body. Eumelanin is a dark brown pigment composed of an aggregate of small subunits (polymers). Eumelanin imparts varying shades of brown to hair and skin, with high concentrations found in the skin of darkly pigmented people.

FACULTATIVE PIGMENTATION: The deepened skin coloration, or tan, that results from exposure of skin to sunlight, as UVR causes the activation of melanin production in melanocytes.

FIBROBLASTS: Connective tissue cells. Fibroblasts that differentiate into cartilage-producing cells, collagen-producing cells, and bone-producing cells form the fibrous tissues of the body.

FOLATE: The vitamin of the B vitamin group that is necessary for DNA replication and cell division. The synthetic form of the vitamin is called folic acid.

FREE RADICALS: Short-lived, highly reactive molecules that have one or more unpaired electrons and are common by-products of normal chemical reactions occurring in cells. Among the most common are the superoxide and hydrogen peroxide ions. In scientific literature, they are often called reactive oxygen species. Free radicals are produced by UVR and can damage DNA.

GENOMICS: The study of the sequence and activity of genes. Comparative genomics involves comparing gene sequences and patterns of gene activation of various organisms.

GENOTYPE: The genetic composition of an individual.

HAPTIC: Referring to the sense of touch.

HOMINID: In common usage, a human ancestor who routinely walked on two legs—that is, who was bipedal. In scientific contexts, bipedal human ancestors are classified as part of the zoological subfamily Homininae and are referred to as hominins.

IMMIGRANT CELLS: Cells that migrate into the skin from elsewhere in the body during early development and that retain some ability to move out of the skin during life. Melanocytes and Langerhans cells are two of the most important types of immigrant cells in the skin.

INTEGUMENT: The covering of the body, including the skin, hair, nails, and scales.

KELOID: An enlarged scar caused by an overproduction of collagen in the process of wound healing. Keloid scars occur more frequently in people with darkly pigmented skin.

KERATIN: A tough, fibrous, insoluble protein found in the outer layer of the skin (epidermis) of humans and other land-living vertebrates. Two major types of keratin exist, alpha (α) and beta (β), with alpha found in mammals and beta found in birds, amphibians, and reptiles. Keratin also makes up most of human hair and nails. Because of its ubiquity and utility, keratin is sometimes referred to as "nature's plastic."

KERATINOCYTES: The main type of cell found in the epidermis. The skin gets much of its color from the packages of melanin pigment (melanosomes) that are injected into these cells from melanocytes.

LANGERHANS CELLS: Spidery (dendritic) cells of the immune system located in the stratum spinosum of the epidermis. They are a type of immigrant cell, moving into the skin from the bone marrow during early development. Known as the "sentinels" of the skin, Langerhans cells are part of the body's first line of defense against invading microorganisms. Langerhans cells transport antigens (foreign proteins) to lymph nodes, stimulating the immune system to produce infection-fighting lymphocytes.

MELANIN: The dense pigment that imparts most of the color to skin. The primary type of melanin in human skin is the very dark eumelanin; reddish-yellow pheomelanin is present in much smaller amounts.

MELANOCYTES: Melanin-producing cells in the skin.

NEW WORLD: The Western Hemisphere, comprising the continents of North and South America.

OLD WORLD: The Eastern Hemisphere, comprising Europe, Asia, and Africa.

ORTHOLOGS: Genes from different species that evolved from the same ancestral gene. Orthologs usually have the same function in the various species.

OSSIFICATION: An area of bone formation. Ossifications in the skin of vertebrates are called dermal bones or dermal ossifications.

PERINEAL: Pertaining to the perineum, the area on the posterior end of the body where the external genitalia and anus reside.

PHENOTYPE: The observable structural and functional properties of an organism that result from the interaction between the organism's genetic composition and its environment.

PHEOMELANIN: The reddish-yellow type of melanin found in the hair and skin of lightly pigmented people. Like eumelanin, its darker counterpart, pheomelanin is a polymeric pigment composed of multiple smaller subunits. Unlike eumelanin, pheomelanin generates rather than neutralizes free radicals when it is exposed to UVR and thus may be implicated in the development of skin cancer in fair-skinned people.

PHOTOPROTECTION: In the context of skin research, protection of the skin from ultraviolet radiation (UVR).

RICKETS: A vitamin D deficiency disease in children, more formally referred to as nutritional rickets. It is characterized by weak and poorly mineralized bones that are easily deformed by the body's weight.

SEXUAL DIMORPHISM: A genetically determined difference between the sexes in an anatomical characteristic such as overall body size, tooth size, or skin color.

SKIN PHOTOTYPE: The classification of skin according to its reaction to UVR, ranging from I (lightly pigmented and never tans) to VI (darkly pigmented and tans profusely).

STRATUM CORNEUM: The topmost layer of the epidermis, composed of a keratin-rich epithelium that protects against abrasion and absorption of large amounts of water or other molecules. The stratum corneum is in a constant state of renewal.

TETRAPOD: The evolutionary lineage of vertebrates possessing four legs with toes. This group contains amphibians, reptiles, birds, and mammals.

THERMOREGULATION: Regulation of body temperature, a phenomenon of particular importance to homeothermic (warm-blooded) animals.

TROPICS: The area between the Tropic of Cancer in the Northern Hemisphere (23.5° or 23°26' N) and the Tropic of Capricorn in the Southern Hemisphere (23.5° or 23°26' S), including the equator. This area experiences little seasonal change during the year because the sun is always high in the sky.

UVA: Ultraviolet radiation of relatively low energy, spanning wavelengths from 315 to 400 nanometers.

UVB: Ultraviolet radiation of relatively high energy, spanning wavelengths from 280 to 315 nanometers.

UVC: The highest-energy ultraviolet radiation, spanning wavelengths from 100 to 280 nanometers.

UVR: Ultraviolet radiation, the type of solar radiation that is of shorter wavelength and therefore greater energy than visible light. UVR comprises the range of wavelengths from 100 to 400 nanometers.

VITAMIN D: An essential vitamin that is responsible for the absorption of dietary calcium from the digestive system and is necessary for the growth and strength of bones. The manufacture of vitamin D begins from cholesterol precursors in the skin, triggered by UVB in sunlight. The bioactive form of the vitamin, vitamin D_3, is produced via a series of chemical transformations that take place in the liver and kidney after the process is initiated in the skin.

NOTES

Introduction

1. Richardson 2003.
2. The beginning of the human lineage—that is, the point at which the lineage leading to humans diverged from that leading to modern chimpanzees—is dated at approximately 6 million years ago. This estimate is supported by bodies of both molecular and paleontological data. The molecular evidence comes primarily from studies of the differences in the nucleotide sequences of nuclear and mitochondrial DNA among humans, chimpanzees, and other primates. As to fossil evidence, the recent discovery and description of the fossil species *Sahelanthropus tschadensis,* from Chad, provide the earliest possible paleontological bookmark of this divergence, at an estimated 6 million years ago. If this species proves to be more closely related to chimpanzees (or other apes) than humans, the earliest known unequivocal member of the human lineage would then be the species *Australopithecus anamensis,* from deposits in the Lake Turkana Basin of Kenya, dated at 4.4 million years ago.

3. The skin, which accounts for about 15 percent of total body weight, has long been considered the body's largest organ, with a surface area measuring one and a half to two square meters (approximately fifteen to twenty square feet) in adults. Some authorities have challenged this description, noting that the mucosal lining of the intestines is actually larger in surface area than the skin and that the total mass of skeletal muscle in the body is heavier. Nonetheless, the skin is certainly one of the body's largest organs, however one measures it.
4. The loss of body hair is one of the most important but least recognized innovations that occurred during human evolution. Among the many studies of this topic, Peter Wheeler's analyses, which demonstrate the benefits of hairless, sweating skin in maintaining normal body temperatures under hot environmental conditions, are the most rigorous (see Wheeler 1988). Wheeler's thermoregulatory model of the origin of bipedalism is incorrect (Chaplin, Jablonski, and Cable 1994), but his recognition of the role of body cooling through sweating in human evolution remains of signal importance.
5. Di Folco 2004, 8.
6. Recent monographs by Claudia Benthien (2002) and Steven Connor (2004) focus on the symbolism connected with skin in art and literature. These works provide many vivid examples of how "skin" as a word and skin imagery have been used to convey notions of humanity, identity, vulnerability, and estrangement, among other things.
7. "Whispers of Immortality" is included in Eliot 1920.
8. Groning 1997; Di Folco 2004; Polhemus 2004.

Chapter 1. Skin Laid Bare

1. The significant anatomical differences between the skin of humans and the skin of our hairier mammalian relatives were first discussed in detail in a series of landmark papers by skin biologist William Montagna, who brought attention to the great importance of the enhanced elasticity, abrasion resistance, and sweating ability of human skin. See, e.g., Montagna 1981.
2. The epidermis varies in thickness from 0.4 to 1.5 millimeters (0.016 to 0.059 inches) over the surface of the body (Chu et al. 2003).

3. For a discussion of various types of oxidative stress, see Elias, Feingold, and Fluhr 2003.
4. The thickening of the stratum corneum caused by repeated UVR exposure is most pronounced in people with darkly pigmented or heavily tanned skin (Taylor 2002).
5. Peter Elias and colleagues have researched the physical and biochemical properties of the epidermis in detail, producing many authoritative papers and book chapters on the subject. Readers seeking more details should consult at least one of these references, e.g., Elias, Feingold, and Fluhr 2003.
6. Chimpanzee Sequencing and Analysis Consortium 2005. Montagna long held that the evolution of a highly effective epidermal barrier function in human skin was one of its true singularities. The sequencing of the chimpanzee genome and the discovery of the unique functional complex that distinguishes human from chimpanzee skin vindicates one of his most important conjectures.
7. In most people, the total thickness of skin ranges from 1.5 to 4.0 millimeters (0.059 to 0.157 inches) (Chu et al. 2003).
8. Smooth leather consists of both the tanned epidermis and the dermis, whereas suede consists of skin that has had some or all of the epidermis removed. Parchment is untanned skin that is derived mostly from the skins of calves, sheep, and goats and that has been dried under tension. Rawhide, as its name implies, is also untanned.
9. In the active field of evolutionary developmental biology ("evo-devo"), study of the evolution of skin in vertebrates is now producing new insights into the genetic foundations for the major features of the integument, including the stratum corneum and the appendages (notably hair and feathers). For a comprehensive and understandable review of the subject, see Wu et al. 2004. Recent research has shown that hair follicles and feather follicles arose independently in evolution around 225 to 155 million years ago, although both share the ability to undergo essential cyclic regeneration (Yue et al. 2005).
10. Charles Darwin was one of the first to recognize the significance of human facial expressions and to draw parallels between the expressions of humans and those of other animals. His work on the evolution and social significance

of facial expressions in humans has been carried on by Paul Ekman and his students, whose inspired and compendious research forms the basis for our modern understanding of the topic (Ekman 1998, 2003).

Chapter 2. History

1. The size, position, and rugosity (roughness) of muscle attachment areas on the bones of fossil animals allow paleontologists to reconstruct significant functional details of the muscle that was once attached there, including the muscle's probable size and range of movement and an assessment of its importance relative to others that were attached to the same bones. This information gives paleontologists insights into the ways ancient animals may have chewed and what they may have eaten. When information on muscle attachments on long bones is combined with data on the dimensions of limb bones, researchers can reconstruct many details of an animal's mode and speed of locomotion.
2. A delightful exception was the discovery in February 2000 of the tracks of the dinosaur *Dilophosaurus* on the Johnson farm near St. George, Utah. The tracks of this 200-million-year-old reptile were made in fine clay, which preserved the impressions of the animal's scaly skin. More information is available online at http://scienceviews.com/dinosaurs/dinotracks.html.
3. Chiappe et al. 1998.
4. Anthropologist Donald Brothwell presents a clear picture of the conditions under which bodies can be preserved for long periods; see Brothwell 1987.
5. The mummified remains of several individuals, including the famed "Beauty of Loulan," were recovered from shallow graves at Qawrighul and other sites in the Xinjiang Uyghur Autonomous Region of China. These mummies attest to the presence of people of central Asian stock in western China in the mid-Holocene. Their cold-weather-adapted clothing resembles that of another famed find, Ötzi, the so-called Neolithic Iceman, who was retrieved from a thawing glacier on the alpine border of Austria and Italy (Barber 2002).
6. For a popular yet comprehensive account of the discovery, recovery, and scientific study of Ötzi, see Fowler 2000.
7. Ding, Woo, and Chisholm 2004.

8. Whitear 1977.
9. For a concise and authoritative discussion of the function and formation of keratins in vertebrate skin, see Spearman 1977. Ping Wu and colleagues provide a more detailed and updated account of the evolutionary developmental biology of keratins that includes extensive discussion of the gene complexes associated with their production in different vertebrates (Wu et al. 2004).
10. Amphibians produce four major classes of noxious compounds in the granular glands of their skin, some of which have been demonstrated to have medical and pharmaceutical uses (Clarke 1997).
11. Wu et al. 2004.
12. The dermal ossifications of reptiles are the antecedents of dermal bones in mammals. Mammal skeletons consist of two major types of bones, those that form by the replacement of cartilage and those that are induced directly in the skin. The latter, dermal bones, form important elements of the mammalian skull, including the bones of the top of the braincase and the lower jaw. It is common in evolution for structures that evolve originally for one purpose—in this case, dermal bones evolving for protection against abrasion in crocodilians—to be co-opted for another. The term "exaptation" was coined to describe this phenomenon. The dermal bones of the mammalian braincase do protect the brain but no longer serve as tough, external armor as they did in reptilian ancestors.
13. Wu and colleagues present a detailed summary of the genetic evidence for the evolution of feathers; see Wu et al. 2004. The historical sequence of feather evolution as seen in the fossil record is reviewed in two excellent papers: see Chiappe 1995; Chuong et al. 2003.
14. Padian 2001.
15. Carpenter, Davies, and Lucey 2000.
16. C. P. Luck and P. G. Wright (1964) describe transepidermal water loss in the hippopotamus. Saito Saikawa and colleagues recently characterized the chemical properties of the hippo's "red sweat"; see Saikawa et al. 2004. On the evolutionary and paleoenvironmental implications of the properties of hippo skin, see Jablonski 2004.

17. Jablonski 2004.
18. In the technical terms of comparative biology, reconstructing the history of adaptation in a lineage requires first that we have a strong working hypothesis of the phylogenetic relationships between the animals of interest. We then need to know how the skin functions in humans as well as in our close relatives so that we can understand what functional transitions must have occurred. How do animals compare, for instance, in their sweating rates and their levels of pigmentation? When these requirements are met, it then becomes possible to reconstruct the major steps that occurred in the evolution of human skin. This methodology is embraced by the field of historical morphology, whose goal is reconstructing the history of adaptations using well-established phylogenies as foundations for retracing the steps of structural and functional transformations that occurred through time.
19. The evolutionary relationships among these animals was a subject of great controversy in biology until the 1980s, when molecular evidence (provided by the sequence of nucleotides in mitochondrial and nuclear DNA) clearly demonstrated this sequence of branching events. Anatomical evidence has been the traditional method of reconstructing evolutionary relationships, but problems in many groups of organisms with convergent evolution—the evolution of similar structures in response to similar environmental influences—can make the reconstruction of evolution relationships difficult and unreliable. The most robust phylogenies combine both types of evidence.
20. Ruvolo 1997.
21. Marks 2003. Molecular evolutionary studies of the past thirty years have shown that many of the "major" morphological differences between humans and chimpanzees may be attributed to relatively few genetic changes (Khaitovich et al. 2005).
22. The three main properties of the skin of Old World anthropoid primates discussed here—hairiness, the ability to produce sweat, and the ability to produce melanin pigment—are shared to a greater or lesser extent with all mammals.
23. Jablonski and Chaplin 2000.

Chapter 3. Sweat

1. The aquatic ape hypothesis was introduced by Alistair Hardy (1960) and subsequently elaborated by Elaine Morgan (1982).
2. In this book, I use the term "hominid" to denote an extinct member of the lineage to which modern humans belong. This is the lineage of primates that separated from that leading to modern chimpanzees about six to seven million years ago. Hominids include many extinct species, some that are closely related and others only distantly related to modern humans. In recent years, the word "hominin" has been used in the scientific literature to refer to the same group, but this term is not as widely understood.
3. During the 1960s and early 1970s, the scholarly literature on human evolution was dominated by works that emphasized the importance of hunting in early prehistory. This inspired some writers to create popular works on human evolution that emphasized the importance of instinctual hostile aggression and killing in shaping human psychology and the direction of prehistory. The most infamous of these was *African Genesis*, by amateur anthropologist Robert Ardrey (1961). The enormous popularity of the aquatic ape hypothesis was a reaction against Ardrey's view of human nature.
4. In addition to diseases that involve waterborne parasites burrowing into the body, malaria—caused by a parasite that infects a water-breeding mosquito—leads to even greater morbidity and mortality throughout the tropics.
5. The idea that hominids may have utilized shellfish and other shallow-water food resources is currently being pursued by Alan Shabel, a PhD student in the Department of Integrative Biology at the University of California at Berkeley.
6. Wheeler 1985.
7. The advocates of the aquatic ape hypothesis have also opined that habitual bipedal posture and locomotion originated while humans were partially aquatic. Many other more convincing hypotheses that are consistent with available facts have been developed to account for this key innovation in human history. George Chaplin and I have advocated a hypothesis for the origin of bipedalism based on bipedal social displays as efficacious methods of social control (Jablonski and Chaplin 1993).

8. Pagel and Bodmer 2003.
9. Bar-Yosef 2002.
10. In fact, it is possible that, if anything, humans actually became hairier when they started to wear clothing routinely. The indigenous peoples of Europe are hairier than others, almost certainly as the result of rehairing, the secondary gain of body hair following its original loss in early hominid history. The regaining of some body hair may have occurred in European populations as an adaptation to prevent skin chafing and subsequent infection, which might have been caused by constantly wearing heavy clothes made from wool or plant fibers.
11. Folk and Semken 1991.
12. The importance of sweating in human evolution was most convincingly argued in a series of papers by Peter Wheeler in the 1980s and early 1990s. This research demonstrated that a functionally naked skin with efficient sweating capabilities was essential for hominids engaged in foraging and other heat-producing activities in the hot, open environments where early members of the genus *Homo* lived (Wheeler 1984, 1985, 1991a, 1991b; Zihlman and Cohn 1988; Chaplin, Jablonski, and Cable 1994).
13. The two main types of sweating in humans are thermal sweating and emotional sweating. They are stimulated by different sets of nerves and occur under different sets of circumstances. Chapter 8 discusses emotional sweating in detail.
14. In particular, such mammals have developed mechanisms for keeping the most heat-sensitive organ, the brain, cool. These include a cross-current heat conduction system at the base of the brain, the so-called *rete mirabile,* which transmits cooled venous blood from the nose directly to veins at the bottom of the brain as the blood passes to the heart. This mechanism is common in large hooved mammals such as deer, buffalo, and antelope. The same effect (circulating cooled blood to the base of the brain) is achieved by panting: evaporation from most surfaces in the mouth cools the blood in the veins of the tongue and cheeks, which is then transmitted to the base of the brain before returning to the heart. This mode of cooling is common in carnivores such as wolves, dogs, and lions.

15. Nonhuman primates possess eccrine sweat glands like those of humans, but these glands usually function minimally and are outnumbered by apocrine sweat glands in most species. Only in a few Old World anthropoid primates, such as the patas monkey, do we find significant eccrine sweating abilities. Patas monkeys are some of the few nonhuman primates found in very open country in equatorial Africa, where they live in small groups and run long distances to find good foraging spots and escape predators. Patas monkeys are the fastest nonhuman primates known. For them, keeping cool during bouts of strenuous exercise under the equatorial sun has meant evolving increased eccrine sweating. When we examine the sweating ability of patas monkeys in the laboratory, we find that they have a preponderance of eccrine sweat glands and that, using these glands, the animals are capable of prodigious sweating compared to even their close monkey cousins (Mahoney 1980).
16. There is no clear consensus about which species name should be applied to the earliest members of the genus *Homo* in Africa. Most scientists would agree that the earliest species to show modern human limb proportions and activity levels was *Homo ergaster.*
17. The rise in activity levels among members of the genus *Homo* has been inferred from studies of fossil skeletons and modern physiology. The average brain size in some of the earliest members of the human lineage such as *Australopithecus afarensis* was 450 cubic centimeters, compared to the approximately 700 or 750 cubic centimeters observed in early *Homo.*
18. Most investigators consider the partial skeleton nicknamed Turkana Boy (KNM-WT 15000) to be a member of the species *Homo ergaster.* Although this species designation is not universally accepted, no one contests the assignment of the specimen to the genus *Homo*. Detailed study of the skeleton has yielded a wealth of information, including estimates of the individual's age, health status, and diet (Walker and Leakey 1993).
19. Studies examining the distance between finds of stone tools, stone tool "workshops," and the natural location of stone for fabricating tools have shown that early *Homo* traveled up to twenty kilometers (about twelve and a half miles) in search of appropriate materials. This information demonstrates not only that hominids at that time had the physical ability to walk

long distances but also that they had the ability to navigate to and from distant sites.

20. Comparative studies of the limb proportions and joint dimensions of early *Homo* have long indicated that these early members of our own genus were anatomically equipped for long-distance walking and running. These studies have now been augmented by a recent experimental demonstration showing that we are maximally efficient during running exercise (Ruff 1991; Carrier 1984; Bramble and Lieberman 2004).
21. Jerison 1978, 1997.
22. Dean Falk (1990) likens the veins around the brain to the radiator of a car that works to keep the engine cool. The susceptibility of the brain to heat stress and the importance of sweating in maintaining the brain within a small temperature range have been the subject of many experimental studies, notably those of Michael Cabanac and colleagues (Cabanac and Massonnet 1977; Cabanac and Caputa 1979; Caputa and Cabanac 1988). For a clear discussion of the importance of whole-body cooling to human health, see Nelson and Nunneley 1998.
23. Wheeler advanced this explanation as a hypothesis for the evolution of bipedalism. Although Chaplin and colleagues have proven this theory incorrect, a small thermal advantage of bipedalism over quadrupedalism would permit bipeds, who had already evolved good sweating capabilities, to be active during the hours of the day when the sun was directly overhead (Wheeler 1984; Chaplin, Jablonski, and Cable 1994).
24. Zihlman and Cohn 1988; Folk and Semken 1991; Goldsmith 2003.
25. Morbeck, Zihlman, and Galloway 1993; Folk and Semken 1991.
26. Knip 1977.
27. Roberto Frisancho's excellent book on human adaptation and accommodation (1995) offers an authoritative yet highly readable account of the body's thermoregulatory reactions and acclimation to heat stress.
28. Zihlman and Cohn 1988.
29. Pandolf 1992.
30. This observation is in accord with Fourier's Law of Heat Flow, which states that the rate of heat loss is directly proportional to the difference between the tem-

perature of the body and that of the environment, and inversely proportional to the thickness of the body shell. Therefore, the greater the surface area, the greater the rate at which core heat can be dissipated. Conversely, the thinner the barrier between the core and the exterior, the greater the rate of heat flow.

Chapter 4. Skin and Sun

1. The World Health Organization, a United Nations agency, maintains an excellent Web site with information on UVR and human health, available at www.who.int/uv/uv_and_health/en/. Another valuable resource is the Web site of the New Zealand Dermatological Society Incorporated, available at http://dermnetnz.org/site-age-specific/UV-index.html. Lynn Rothschild (1999) offers a persuasive discussion, illustrated with striking examples, of the importance of UVR as a creative force in evolution.
2. This map was created by George Chaplin from data collected by the NASA Total Ozone Mapping Spectrometer (TOMS), version 7 (Herman and Celarier 1996), using the procedure described in Jablonski and Chaplin 2000. The NASA TOMS satellites collect data on UVMED, the minimal erythemal (redness-producing) dose of UVR. The UVMED is specifically defined as the amount of UVR necessary to produce a barely perceptible reddening of lightly pigmented skin.
3. Other factors that affect the amount of UVR received at the earth's surface include the season, the moisture content of the local atmosphere, the depth of the ozone column, and orbital parameters (such as the closeness of the earth to the sun at a particular time based on variation in the earth's orbit) (Hitchcock 2001; Madronich et al. 1998).
4. Caldwell et al. 1998; Johnson, Mo, and Green 1976.
5. The mechanisms of how UVR causes DNA damage to human skin have been widely studied and are now well known. James Cleaver has carried out or directed some of the most authoritative such studies; his research has centered on patients suffering from a rare genetic disease, xeroderma pigmentosum, in which damaged DNA cannot repair itself. Cleaver and his colleague Eileen Crowley recently published an excellent review of UV damage, DNA repair, and skin cancer (Cleaver and Crowley 2002). Other, even more recent reviews

of this complex and fascinating subject are now available: see Kappes et al. 2006; Pfeifer, You, and Besaratinia 2005.

The most common photoproducts produced by UVR acting on DNA are cyclobutane pyrimidine dimers (CPDs).

6. For a recent review of the differences in the nature of the DNA damage produced by UVA and UVB, see Pfeifer, You, and Besaratinia 2005. The possibility that UVA plays a key role in the etiology of malignant melanoma has recently been brought to light: see Garland, Garland, and Gorham 2003; Matsumura and Ananthawamy 2004.
7. Cosentino, Pakyz, and Fried 1990; Mathur, Datta, and Mathur 1977.
8. The subtle influence of folate on the cell's reproductive capacity has led to the recognition that even marginal folate deficiencies may have significance in developmental disorders and degenerative diseases associated with high morbidity and mortality (Lucock et al. 2003).
9. Bower and Stanley 1989; Fleming and Copp 1998; Suh, Herbig, and Stover 2001.
10. Many reliable and understandable sources of information on the relevance of folate levels to health are now available on the Internet; see, for example, http://ods.od.nih.gov/factsheets/folate.asp.
11. The light-induced decomposition (photolysis) of folate has been experimentally demonstrated at 340 nm and 312 nm, in the UVA and near-UVA wavelengths, as well as by gamma radiation (Hirakawa et al. 2002; Kesavan et al. 2003; Lucock et al. 2003; Off et al. 2005). Nanometers (abbreviated as nm) are metric units of length used to measure the wavelengths of electromagnetic radiation.
12. Richard Branda and John Eaton (1978) demonstrated that folate undergoes photolysis in vitro when subjected to UVA (at a strength of 360 nm) and that serum folate levels of human subjects dramatically declined after long-term exposure to the same wavelength (thirty to sixty minutes once or twice a week for a minimum of three months). These researchers and others (e.g., Zihlman and Cohn 1988) intimated that this process had relevance to the evolution of skin color but did not implicate a specific causal mechanism.
13. Off et al. 2005.

14. Jablonski and Chaplin 2000.
15. For mammals, it is probably better to consider vitamin D as a hormone rather than as a vitamin because it is derived from a cholesterol-like precursor in the skin, 7-dehydrocholesterol (Holick 2003).
16. Michael Holick has pioneered research on the chemistry, biological activity, and clinical importance of vitamin D. His interest in the distribution of vitamin D in different life forms has led him to include important evolutionary insights in many of his papers (e.g., Holick 1995, 2003).
17. Studies of the production and mode of action of vitamin D in the laboratory of Michael Holick at Boston University Medical School have resulted in identification of the wavelengths of UVR that initiate vitamin D production in the skin (MacLaughlin, Anderson, and Holick 1982); chemical characterization of the precursor and previtamin D molecules (Webb, Kline, and Holick 1988); and definitive identification of the final, biologically active form of the vitamin as 1α, 25-dihydroxyvitamin D_3. For many years, some theorized that excessive sun exposure could cause the body to produce too much of the active form of vitamin D, leading to vitamin D poisoning, or "vitamin D intoxication." This mechanism was invoked as the reason dark skin evolved in the tropics (Loomis 1967). Subsequently, however, researchers have shown that overproduction of the active form of vitamin D is impossible (Holick, MacLaughlin, and Doppelt 1981).
18. Wharton and Bishop 2003; Holick 2001; Yee et al. 2005.
19. Holick 2001; Yee et al. 2005.
20. Garland et al. 2006; Grant 2003.

Chapter 5. Skin's Dark Secret

1. Ortonne 2002; Sulaimon and Kitchell 2003.
2. Shosuke Ito (2003) presents a readable and endearing account of the struggle to determine the chemical formula of eumelanin, one of two types of melanin found in the skin of humans and other mammals.
3. Kollias et al. 1991; Ortonne 2002.
4. Kollias 1995a; Sarna and Swartz 1998.
5. Kaidbey et al. 1979; Kollias 1995a, 1995b.

6. Young 1997.
7. This elegant study showed that the variant form of the gene (slc24a5) that was common in golden zebrafish was the ortholog, or functionally comparable form, of the predominant gene in European people. The evolutionarily conserved and ancestral form of the gene is found in native African and Asian populations (Lamason et al. 2005).
8. Pheomelanin imparts the red color to red hair and the reddish cast to freckles in fair skin. It also appears to be present in the skin of East Asians and their descendant populations in the Americas (Thody et al. 1991; Alaluf et al. 2002). The presence of pheomelanin in the skin of some Asian peoples led to them being labeled "yellow-skinned."
9. Ortonne 2002.
10. Other common anomalies of pigmentation are piebald spotting and various types of hyperpigmentation (Sulaimon and Kitchell 2003; Robins 1991).
11. In his comprehensive book-length review of human pigmentation, Ashley Robins (1991) provides a concise treatment of the biochemical basis and manifestations of albinism in humans.
12. Research on the density and activity of melanocytes in humans has revealed that the number of pigment-producing cells in the body is remarkably uniform among living humans, but that the activity of these cells varies according to genetically determined levels of pigmentation, UVR exposure, and age (Fitzpatrick, Seiji, and McGugan 1961; Halaban, Hebert, and Fisher 2003; Lock-Andersen, Knudstorp, and Wulf 1998).
13. Several excellent papers now provide details of the production and activity of free radicals in skin; see Ortonne 2002; Sulaimon and Kitchell 2003; Young and Sheehan 2001.
14. On the importance of protecting DNA from UVR-induced damage and the role that melanin plays in this effort, see Cleaver and Crowley 2002. For a discussion of the destruction of folate by UVR and other very high-energy forms of radiation, see Kesavan et al. 2003.
15. Thomas Fitzpatrick and Jean-Paul Ortonne (2003) have recently published an authoritative treatment of the tanning response. The relationship between constitutive pigmentation and susceptibility to skin cancer has been the sub-

ject of numerous studies, including Sturm 2002 and Wagner et al. 2002, which deal specifically with the increased risk of sunburning and skin cancer in genetically lighter-skinned people.

16. Olivier 1960; von Luschan 1897.
17. Joseph Weiner was the first to describe the usefulness of the EEL reflectance spectrophotometer (manufactured by Evans Electroselenium Ltd.) in the measurement of human skin color. Weiner was a distinguished British anthropologist who is also credited with exposing the Piltdown hoax. (The so-called Piltdown man was discovered in a gravel pit in England by Charles Dawson in 1912. Heralded at first as a "missing link," Piltdown man, known by the scientific name *Eoanthropus dawsoni*, was a sensation because it boasted anatomical features of both modern humans and apes. In 1953, a team of scientists led by Weiner demonstrated that Piltdown man was a hoax, constructed from a medieval human skull and an orangutan's lower jaw. The identity of the perpetrator of the hoax remains a source of controversy in paleoanthropology.)
18. Wassermann 1974.
19. Fitzpatrick and Ortonne 2003.

Chapter 6. Color

1. We can determine the ancestral condition for a specific feature of a lineage in a variety of ways. The most common method is outgroup analysis, which uses the state in which a trait exists in a closely related organism to infer the ancestral state of the feature in the lineage being studied. With the traits of skin color and sweating ability in apes, we can look to the next most closely related group—Old World monkeys—as the outgroup from which we can determine the likely ancestral conditions. For all the catarrhine primates, the group comprising Old World monkeys, apes, and humans, we can infer the ancestral condition by looking at the next most closely related group, the platyrrhines, or New World monkeys, and so forth. This mode of analysis shows that the ancestral condition for all primates includes lightly pigmented skin, predominantly apocrine sweat glands, and a covering of dark hair.
2. Unlike the skin of other catarrhines, the skin of gorillas—the African apes

most closely related to both chimpanzees and humans—is darkly pigmented and covered with dark hair. Although gorillas dwell in African tropical forests, they spend much of the daytime foraging in open glades or swamps within the forests. These openings are located in low-lying sites or in places where large trees have fallen. These "salad bowls" within the forest provide a wealth of vegetation on which the gorillas can feed. Unlike the surrounding forest, however, these openings are exposed to the full measure of equatorial sunlight and UVR. Because gorillas spend long hours foraging in such places, it is likely that natural selection favored the evolution of protective dark pigmentation.

3. Two other putative hominids have been recognized: one emerged before *Australopithecus,* and the other is dated to about the same time. These are *Sahelanthropus tschadensis,* an approximately 6-million-year-old form from Chad; and *Ardipithecus ramidus,* a species from Ethiopia estimated to be 4.5 million years old. The status of these species cannot yet be conclusively judged because the details of their anatomy, posture, and locomotion are still unclear. Bipedalism is considered the key defining adaptation of the hominid lineage.
4. The name "australopithecine" is widely used to designate all the hominids that existed before the emergence of *Homo,* including species of the genera *Australopithecus, Paranthropus,* and *Kenyanthropus.* The genus *Australopithecus* comprises at least four different species: *A. anamensis* (known from deposits about 4.4 million years old) from Kenya; *A. bahrelghazali* (4 million years old) from Chad; *A. afarensis* (3.6 to 3.2 million years old) from Ethiopia and Tanzania; and *A. africanus* (about 3 million years old) from South Africa. Closely related are the so-called robust australopithecines, which exhibit larger teeth and jaw structures than those known as gracile australopithecines and appear to have occupied a different ecological niche. The robust australopithecine lineage is generally considered to belong to a separate genus, *Paranthropus,* which diverged from the *Australopithecus* lineage more than 3 million years ago. *Paranthropus* itself comprises three widely recognized species, *P. aethiopicus, P. boisei,* and *P. robustus.* The most recent addition to the australopithecines is the species *Kenyanthropus platyops,* described from deposits

approximately 3.5 million years old from Lomekwi, in the Lake Turkana district of northern Kenya. *Kenyanthropus* exhibits craniofacial characteristics similar to those of acknowledged early members of the genus *Homo,* but it is much older.

5. Over the past thirty years, many australopithecine species have been proposed as direct ancestors of the genus *Homo.* A major problem in choosing among these prospects has been the lack of clarity about the anatomical definition of *Homo* itself. Additionally, some nominees are deemed to have characteristics of their teeth and cranial bones that are too specialized to have given rise to the more generalized anatomical conditions in *Homo.* Generally, chronologically older species are more likely to display generalized anatomical conditions that make them better candidates as direct ancestors of *Homo.* Many paleoanthropologists think that, of all the australopithecine species now recognized, the species *A. anamensis* has characteristics most consistent with those expected of an ancestor for the *Homo* lineage.
6. Most students of human evolution would assign the earliest known examples of the genus *Homo* to the species *Homo ergaster.*
7. Rogers, Iltis, and Wooding 2003.
8. The noted American physical anthropologist Paul Baker and his colleagues conducted the primary research on heat loads in humans and how they vary by skin color (and other factors). Many of these studies were carried out in conjunction with the U.S. military, in order to assess differences in endurance and physiological stress among soldiers in the field who had differing physical appearances and physiques (see, e.g., Baker 1958; see also Daniels 1964). For a review of the general importance of integumentary color in determining heat loads in vertebrates, see Walsberg 1988.
9. The earliest fossil evidence of *Homo sapiens* is described in White et al. 2003. The exact timing of the first exodus of modern people from Africa is still the subject of some debate because of a dearth of relevant fossils. For a review of the pertinent fossil evidence, see Stringer 2003.
10. These ages are based on well-dated archaeological and paleontological sequences, as well as good estimates of population divergence times derived from comparison of DNA sequences. Several authoritative papers have been

published on this subject recently; see, e.g., Underhill et al. 2000; Henshilwood et al. 2002; Adcock et al. 2001; Klein et al. 2004; Luis et al. 2004.

11. The full details of our theory of the evolution of human skin coloration are presented in Jablonski and Chaplin 2000. A detailed summary is also available; see Jablonski 2004.
12. Among adverse effects such as sunburn, sun-related skin degeneration, and skin cancers, only malignant melanoma generally affects people of reproductive age, and those numbers are sufficiently low that they are unlikely to have had a significant effect on the course of natural selection (Jablonski and Chaplin 2000). Harold Blum also argued that dark skin pigmentation could not have evolved primarily as adaptive protection against skin cancer because such cancers rarely cause death during peak reproductive years (Blum 1961). Other adaptive explanations for highly melanized skin include its ability to provide effective concealment in dark habitats such as tropical forests (Cowles 1959) or to impart greater resistance to tropical diseases and parasites (Wassermann 1965). These hypotheses, however, fail to demonstrate real or potential increases in reproductive success (Jablonski and Chaplin 2000; Blum 1961).
13. The production of the biologically active form of vitamin D in the body involves many steps, the first of which occur in the skin. The details of this process have been determined largely as the result of the efforts of Michael Holick, who for many years headed a laboratory at Boston University Medical School devoted to the study of vitamin D and health. The authoritative publications of his research group are highly recommended; see Holick, MacLaughlin, and Doppelt 1981; Holick 1987; Webb and Holick 1988; Webb, Kline, and Holick 1988; Holick 1995, 1997, 2004.
14. Kaidbey et al. 1979; Stanzl and Zastrow 1995.
15. The so-called vitamin D hypothesis for the evolution of lightly pigmented skin was first proposed by Frederick Murray (1934) and then developed in detail by W. Farnsworth Loomis (1967). Not all authorities have agreed with this interpretation, however. Anthropologist C. Loring Brace argued that depigmentation of human skin occurred not as the result of active selection for lighter pigmentation but because selective pressure on pigmentary systems

was relieved as humans populated increasingly high latitudes where dark pigmentation was no longer required as a shield against UVR. Brace's structural reduction hypothesis is based on the "probable mutation effect," whereby mutations in the genes controlling melanin pigmentation accumulated, leading to reduced or failed melanin production. Brace proposed that this effect accounts not only for depigmentation of human skin but also for related phenomena such as loss of eyesight and loss of pigment in all skin-related structures in cave-dwelling organisms (Brace 1963).

16. Webb, Kline, and Holick 1988.
17. Jablonski and Chaplin 2000.
18. This information has been developed into a map, which is available in Jablonski and Chaplin 2000.
19. Cornish, Maluleke, and Mhlanga 2000.
20. After decades of intense scientific debate, the question of whether Neandertals contributed significantly to the ancestry of modern Europeans has been settled satisfactorily by DNA evidence retrieved from the bones of Neandertals at four different, geographically widespread sites. Mitochondrial DNA recovered from Neandertal bones is not found in modern humans. This finding does not conclusively eliminate the possibility that Neandertals contributed genes to the modern human gene pool, but it shows that the Neandertal genes, if present in modern human populations in the past, were swamped by those of modern humans or eliminated by genetic drift (Serre et al. 2004).
21. Lee and Lasker 1959.
22. Fair skin, red hair, and freckles are associated with elevated risk of melanoma and nonmelanoma skin cancer because of the variant forms of the *MC1R* gene, which people with these traits carry (Sturm et al. 2003).
23. Ortonne 2002; Kaidbey et al. 1979.
24. The peak UVR that a person can experience is related to his or her location, with the highest levels occurring during the summer. The UV content of sunlight varies by latitude. Most latitudes receive much more long-wavelength UVR (UVA) because UVA can penetrate the atmosphere from any angle. The atmosphere filters out shorter-wavelength UVR (UVB) for most of the year in most places outside the tropics.

25. Barker et al. 1995.
26. Cleaver and Crowley 2002.
27. For useful accounts of the tanning response, see Kaidbey et al. 1979 and Ortonne 2002. For a description of how reactive oxygen species (free radicals) liberated by UVR exposure break down structural proteins in the skin, see Fisher et al. 2002.
28. Contact between the seafaring nations of post-Renaissance Europe and the coastal reaches of Asia, Africa, and the Americas began on a large scale in the fifteenth century. One of the best compilations and interpretations of the impressions recorded by these travelers is provided by Sujata Iyengar, who explores the cultural mythologies surrounding skin color in Britain (Iyengar 2005).
29. Renato Biasutti's work (1959) provided the earliest comprehensive map showing the distribution of skin color among indigenous peoples. Ashley Robins (1991) sheds light on the shortcomings of this map.
30. The map shown in figure 24 is very similar to the one found in Jablonski and Chaplin 2000 and is based on skin reflectance data measured with the Evans Electroselenium Limited (EEL) reflectance spectrophotometer. This apparatus was widely used by anthropologists studying indigenous peoples of the Old World during the latter half of the twentieth century. Unfortunately, anthropologists studying peoples of the New World used a different device (manufactured by the Photovolt Corporation), which does not provide readings that can be readily converted to the same standard as those from the EEL device. In Australia, von Luschan tiles or other methods of color matching were used to record skin color throughout much of the twentieth century, and these data cannot yet be reliably converted to skin reflectance readings taken from spectrophotometers.

    The study by Jablonski and Chaplin (2000) defines an indigenous population as one that has been in its current location since 1500 CE. This date is reasonable with respect to the inauguration of the modern era of European colonization, though it fails to recognize several major movements of human groups within continents that occurred before 1500 (such as the expansion of Bantu-language groups within Africa). These movements, along with European colonization and the increasingly rapid and distant migrations of human populations through time, have fundamentally altered the human landscape that was established in prehistoric times. Human migrations occurring especially since the advent of agriculture have made the interpretation of geographically and biologically significant trends in human phenotypes difficult.

31. Chaplin and Jablonski 1998.
32. Chaplin 2001, 2004.
33. Jablonski and Chaplin 2000; Frost 1988.
34. Frost 1988; Aoki 2002.
35. During late pregnancy and lactation, human females experience a transient loss of up to 10 percent of the calcium and phosphate stores in their own skeleton (Kalkwarf and Specker 2002; Kovacs 2005).
36. Jablonski and Chaplin 2000.
37. Robins (1991) provides a useful summary of the influence of hormones on skin pigmentation and discusses chloasma as it results from both pregnancy and oral contraceptive use.
38. Ibid.
39. Diamond 2005.
40. Chaplin 2001; Johnson, Mo, and Green 1976.
41. When Eskimo-Aleut populations depart from their traditional diets and adopt modern diets consisting largely of processed foods and foods with low levels of vitamin D, they begin to suffer from a high prevalence of vitamin D deficiency diseases, especially rickets, among other health problems (Gessner et al. 1997; Haworth and Dilling 1986; Moffatt 1995).
42. Barsh 1996; Sturm, Teasdale, and Box 2001.
43. Research on the evolution of the *MC1R* gene is now in full swing and is being conducted by several excellent laboratories around the world. The *MC1R* gene is the human equivalent of the *Agouti* gene in mice, which regulates the production of the eumelanin and pheomelanin pigments of the coat. In humans, the synthesis of eumelanin is stimulated by the binding of α-melanotropin (α-melanocyte-stimulating hormone) to the functional *MC1R* receptor expressed on melanocytes (Barsh 1996; Rana et al. 1999; Scott, Suzuki,

and Abdel-Malek 2002). The significance of variant forms (polymorphisms) of the *MC1R* gene has been studied in considerable detail (e.g., Rana et al. 1999; John et al. 2003), especially in connection with skin cancer susceptibility among people with different forms of the gene (Healy et al. 2001; Scott et al. 2002; Smith et al. 1998). An elegant study by Alan Rogers, David Iltis, and Stephen Wooding (2003) established the importance of dark pigmentation (and a concomitant lack of variation in *MC1R* sequences) in the early history of the genus *Homo*.

44. For an excellent review of the significance of variation in the *MC1R* gene, see Sturm, Teasdale, and Box 2001. Keith Cheng's team at Pennsylvania State University, using an experimental zebrafish model, has recently conducted research elucidating the probable genetic basis of lightly pigmented skin among Europeans (Lamason et al. 2005).
45. Jablonski and Chaplin 2000. This interpretation is supported by the discovery of a variant gene associated with depigmented skin that is found in light-skinned European people, but not in light-skinned Asians (Lamason et al. 2005).
46. Race, Ethnicity, and Genetics Working Group 2005.
47. This delicate and important topic has recently been addressed in two noteworthy papers: see Parra, Kittles, and Shriver 2004; Gravlee, Dressler, and Bernard 2005.
48. The incidence of nonmelanoma skin cancer is at least forty-five times higher among the Japanese population in Kauai, Hawaii, than among the Japanese population in Japan, as a result of Kauai's intense UVR levels and the outdoor lifestyle of its inhabitants (Chuang et al. 1995). Like Europeans who have been "transplanted" to sunnier climes, lightly pigmented Asians suffer high skin cancer morbidity because of their skin color phenotype.
49. Garland et al. 2005; Hodgkin et al. 1973.

Chapter 7. Touch

1. The text paraphrases a sentence from an excellent article by paleontologist Geerat Vermeij (1999), who is blind, in which he describes how he uses touch to study details of morphology.

2. Possessing a grasping hallux (big toe) is one of several characteristics that distinguish primates from other mammals. This feature is associated with a highly mobile ankle joint. Other mammals besides primates, such as rodents and insectivores, exhibit a grasping pollex (thumb), although primates have by far the most dexterous and nimble of these appendages.
3. Chu et al. 2003; Dominy 2004.
4. Several new observational and experimental studies have focused on the behavioral and neurological adaptations of persons with long-term blindness. Vermeij, who has been blind since early childhood, provides a delightful account of how he uses his hands in paleontological fieldwork (1999, 217). The cortical adaptations to blindness, including the "recruitment" of parts of the visual cortex by portions of the brain concerned with hearing and touch, are extreme but important examples of neuronal plasticity, the potential of parts of the nervous system to adapt to internal or external inputs (Van Boven et al. 2000; Sathian 2005).
5. See Miller 2005, a news report of John Zook's recent research on the control of bat flight. Zook reported these findings at the 2005 meeting of the Society of Neuroscience but has not yet published this research in a scientific journal.
6. Galton used the branch and end points of epidermal ridges to establish a probabilistic model of fingerprint individuality. Fingerprints, which do not change throughout an individual's life (Roddy and Stosz 1997), are increasingly used in automated matching systems today.
7. Researchers have conclusively demonstrated that dry adhesion by van der Waals forces is the mechanism that geckos use to grip smooth substrates (Autumn et al. 2002). The relationship between the surface density of setae and the degree of adhesion has led to the fabrication of a new class of adhesives, which employ nanofabricated setal tips.
8. Although the importance of vision in early primate evolution has been appreciated for a long time, we have only recently begun to develop a comprehensive understanding of sensory evolution in primates, largely through the innovative and integrative research of Peter Lucas and Nathaniel Dominy. Lucas and Dominy have made strong cases that acute color vision in primates

and highly discriminatory touch evolved as natural selection acted to improve animals' ability to detect high-quality (often red- or yellow-colored) foods such as ripe fruits and young leaves in their environment (Dominy and Lucas 2001; Dominy 2004).

9. See Montagu 1971, 290–291.
10. Horiuchi 2005.
11. Ashley Montagu (1971) provides a detailed discussion of the importance of the process of vaginal birth in humans. The gauntlet of touch that a newborn experiences as the result of both this process and the practice of maternal cuddling and fondling is of great importance to infant survival. Montagu also cites reports indicating that human infants who are deprived of some or all of these experiences as the result of cesarean or premature birth suffer higher rates of respiratory complications and nervous excitability than full-term infants born by vaginal delivery.
12. Both Ashley Montagu (1971) and Tiffany Field (2001) offer extensive documentation of the nature and benefits of prolonged physical contact between mother and newborn.
13. Research on the responses of premature infants to touch and massage is summarized in Field's highly readable book *Touch* (2001).
14. The well-known and, to some, infamous studies of psychologist Harry Harlow vividly demonstrated the short- and long-term consequences of touch deprivation in macaques (Ruppenthal et al. 1976; Harlow and Zimmerman 1958). Harlow and colleagues found not only that deprived infants suffer from life-long anxiety and irritability but also that mothers deprived of contact with their infants do not learn proper mothering skills. The results of these studies have benefited animals in zoos and primate research facilities but have had limited impact on human infant care (Harlow et al. 1966).
15. Harlow and Zimmerman 1958.
16. During the massage of a newborn infant that I observed in rural Nepal, the infant remained calm throughout a long and energetic whole-body massage and fell deeply asleep immediately thereafter.
17. Field 2001.
18. Volunteer "grandparents" who participated in a one-month study of massage

therapy for infants reported benefits for themselves, including improved lifestyle, a better social life, and fewer medical complaints (Field 2001).

19. The importance of touch in mothering has been recognized for a long time. The observations from the German orphanages come from an original and frequently cited study by Elsie Widdowson (1951) that has been summarized in many subsequent textbooks and popular treatises on child-rearing. Montagu (1971) provides a persuasive summary of the literature on "tender loving care" in orphanages and pediatric hospital wards. Robert Sapolsky (2004) discusses stress dwarfism and its relation to mothering practices.
20. People with autism can be hypersensitive to light touch (Field 2001), finding it distressing or annoying. But for some of these individuals, deep touch or pressure applied over a large area of the body can sometimes be soothing. This phenomenon became well known after Oliver Sacks related the story of Temple Grandin, a scientist with autism who constructed her own "squeeze chair" in order to calm herself when she felt agitated (Sacks 1996). Grandin tells her own story in *Thinking in Pictures* (1996), where she also recounts important information about the calming effect that physical pressure can produce in domestic animals. Sapolsky (2004) discusses the physiological sequelae of deep touch and massage and their beneficial effects on children's growth.
21. Studies by research groups headed by Frans de Waal and others now document the efficacy of grooming in building, maintaining, and healing alliances in a range of Old World monkey species and chimpanzees. For an introduction to these excellent bodies of work, see de Waal and van Roosmalen 1979; de Waal 1993.
22. Sapolsky 2004, 2005.
23. Low-ranking mothers in baboon societies are relegated to the periphery of prime feeding areas. As a result, they and their infants are unable to forage on the more nutritious foods cornered by those of higher rank. Jeanne Altmann and Stuart Altmann, as well as others, have extensively studied the fates of these individuals; see Altmann et al. 1977; Silk, Alberts, and Altmann 2003.
24. Suomi 1995.

25. Aurelli and de Waal 2000; de Waal 1993, 1990.
26. My English mother-in-law exclaimed after my wedding, "I've never been kissed so much in my life!" referring to the affectionate mauling she received from many of my mostly Italian American relatives. Some cultures are liberal in the amount of physical affection lavished on strangers in social contexts, while others are far more reserved.
27. Montagu 1971. On a trip to Kenya in 2004, I monitored an exchange of letters and op-ed pieces in the *Nairobi Standard* concerning the growing popularity of baby strollers. All of the many medical and child care authorities contributing to the debate condemned the contraptions as devices of torture that forced children to be separated from their mothers' loving touch.
28. Montagu presents a passionate discussion of this topic (1971, 274–275).
29. Field 2002.
30. Montagu examines this phenomenon in detail, emphasizing that corporal punishment of children played a role in the development of various manifestations of social pathology in Nazi Germany and early twentieth-century England (Montagu 1971, 275–279).
31. It has been said that "the good doctor is a good groomer" (Dr. Lynn Carmichael, quoted in Field 2001). In other words, a good doctor knows when to provide caring touch to his or her patients.
32. Weze et al. 2005; Dillard and Knapp 2005; Butts 2001; Field 2001.
33. Few rigorous studies have been conducted on the effects of touch and massage as a part of caring for the elderly, but there are some reports of positive and encouraging results (Butts 2001; Gleeson and Timmins 2004). Training health care providers in caring touch is important, especially as increasing numbers of the elderly enter care facilities, but such training must be undertaken with great thoughtfulness and with due attention to the legal rights of all involved.

Chapter 8. Emotions, Sex, and Skin

1. The skin contains nerve receptors and chemicals that allow it to respond involuntarily to a variety of stimuli. Some of these reactions are fast; others are slow. It also produces chemicals of its own, including a suite of hormones,

from thyroid hormone to the sex hormones (androgens and estrogens). Thyroid hormone stimulates activity in the skin's fibroblasts and keratinocytes and appears to be essential for hair formation and sebum production. Androgens promote hair growth in the beard, armpits, and groin. Estrogens inhibit hair growth and decrease sebaceous gland function. Changes in hormone levels that occur as the result of normal aging have profound effects on the appearance of the skin. These are especially marked when androgen and estrogen levels are ramped up during puberty and as they decline with middle age. The skin also produces and converts hormones—for example, it can convert the androgen dehydroepiandrosterone (DHEA) into dihydrotestosterone (DHT). For this reason, some scientists have favored calling human skin the body's largest independent peripheral endocrine organ (Zouboulis 2000).

2. Zihlman and Cohn 1988; Folk and Semken 1991.
3. Christian Collet and colleagues discuss changes in the skin's electrical conductance and other effects that reflect sweat gland activity; see Collet et al. 1997. The unreliability of polygraphs is well established (Saxe 1991), a fact that precludes the admission of polygraph evidence in many courts around the world (Ben-Shakhar, Bar-Hillel, and Kremnitzer 2002). The growing field of psychophysiology now uses computerized polygraphs with new types of sensors for collecting physiological data consistent with deception (Yankee 1995).
4. The heightened responsiveness of facial arteries to emotion is well known (Wilkin 1988), but reactions are highly individual (Katsarou-Katsari, Filippou, and Theoharides 1999).
5. Sinha, Lovallo, and Parsons 1992.
6. Montoya, Campos, and Schandry 2005.
7. Drummond 1999; Drummond and Quah 2001.
8. The sympathetic nerve supply for the face is complex and includes fibers that bring about both vasoconstriction and vasodilation (the constriction and dilation of blood vessels). In anger, these two effects appear to compete, with vasodilation overriding vasoconstriction to produce the typical flushed face (Drummond and Lance 1987; Drummond 1999). In blushing, the sympa-

thetic fibers responsible for vasodilation appear to be more active, suggesting subtle differentiation of pathways in the brain that are responsible for the effect. It is possible that, during blushing, vasodilation is unchallenged by vasoconstriction and thus brings about increased blood flow in facial arteries and heightened facial redness.

The graded nature of the blushing response has been documented (Leary et al. 1992). In acutely embarrassing situations, increases in facial redness, cheek temperature, and finger skin conductance are greater than those in other emotionally charged situations, even rage (Shearn et al. 1990). The intensity of the blushing response appears to be related to a person's predisposition to developing rosacea, a skin condition described in chapter 9. The esteemed dermatologist Albert Kligman explores this relationship in vivid terms in a paper based on one of his recent lectures; see Kligman 2004.

9. Bögels and Lamers 2002.
10. Jablonski and Chaplin 1993; Jablonski, Chaplin, and McNamara 2002.
11. Montagna 1971.
12. Domb and Pagel 2001; Dunbar 2001.
13. In primates, males or females transfer from their natal troop to a new one when they reach sexual maturity. This is one of the mechanisms that has evolved to promote genetic exchange and increased genetic variability in small groups that would otherwise be subject to inbreeding. However, Jane Goodall provides dramatic accounts of the brutal attacks that are sometimes launched by resident chimpanzees against others who approach their territory (Goodall 1986). Other primatologists involved in long-term observational studies of chimpanzees have observed and described similar attacks and "raids" by chimpanzees in the wild in the past thirty years (Manson and Wrangham 1991).
14. Matsumoto-Oda 1998.
15. Abramson and Pearsall 1983. Considerable confusion exists in the scientific literature about where blushing stops and flushing starts (or vice versa) (Kligman 2004). Sexual flushing appears to occur because of changes in the function of the lymphatic system brought about by sexual excitation, but the exact mechanism still isn't clear.

Chapter 9. Wear and Tear

1. Connor 2004.
2. Grevelink and Mulliken 2003.
3. The most common moles are known clinically as acquired typical nevomelanocytic nevi or atypical melanocytic nevi; see Tsao and Sober 2003 for greater detail.
4. Steven Connor writes engagingly on the topic of divination of moles and provides extensive documentation from original sources (2004, 96–108).
5. The initial stages in the healing of skin wounds are well understood from a molecular level and are the subject of several useful reviews; see Diegelmann and Evans 2004; Gharaee-Kermani and Phan 2001. The later stages of wound healing and scar formation are understood in less detail but continue to be investigated using animal models, including insects ("Molecular Biology of Wound Healing" 2004).
6. Taylor 2002, 2003; Ketchum, Cohen, and Masters 1974.
7. Molecular studies of the coevolution of arthropods, their vertebrate hosts, and the diseases often borne by the arthropods are shedding light on major, long-standing human afflictions such as malaria, as well as less serious but vexatious problems such as infestations of body and head lice; see, for example, Hartl 2004; Burgess 2004.
8. Ian Burgess (2004) summarizes further details of the evolutionary relationship between lice and humans in his recent review.
9. Robert Sheridan (2003) provides a brief but authoritative summary of burns and their treatment.
10. For a readable treatment of burn care, see Ravage 2004.
11. Many excellent Web sites are available to help people understand the common forms of dermatitis. These include a useful self-diagnosis chart of skin rashes and other changes, compiled by the American Academy of Family Physicians, which can be found online at http://familydoctor.org/545.xml.
12. Ikoma et al. 2003.
13. Gary Fisher and colleagues provide an authoritative summary of the physiological processes involved in skin aging (Fisher et al. 2002). With age, the

skin loses some of its effectiveness as a barrier, and many of its functions decline or become slower, including cell replacement, wound healing, sweat and sebum production, vitamin D production, and DNA repair, among others.

14. The literature on skin cancer is vast. For recent overviews of the causes and epidemiology of skin cancer, see Christenson et al. 2005; de Gruijl and van Kranen 2001; Sturm 2002; Garland, Garland, and Gorham 2003.
15. These estimates are based on figures provided by the American Cancer Society, whose useful Web site on nonmelanoma skin cancers is available at www.cancer.org/docroot/CRI/content/CRI_2_4_1X_What_is_skin_cancer_51.asp?sitearea=.
16. Erb et al. 2005; Cleaver and Crowley 2002.
17. Christenson et al. 2005.
18. Sturm et al. 2003; Newton Bishop and Bishop 2005.
19. Many factors determine the prognosis of melanoma. These include the thickness of the lesion, the degree of ulceration of the primary lesion on the skin, and the amount of lymph node involvement. Factors such as a person's age and the site of the lesion also make a difference in the likely clinical course. In malignant melanoma, the site of the metastases and their associated blood supplies strongly influence the outcome (Homsi et al. 2005).
20. For more information on preventing skin cancer, see the excellent online resources provided by the National Cancer Institute, available at www.cancer.gov/cancertopics/pdq/prevention/skin/Patient/page2.

Chapter 10. Statements

1. Throughout human history, tattoos have been important for identifying those who have died in battle. Even today, official applications for the military and various forms of formal identification include questions about any permanent marks or tattoos on the body that might be useful in establishing a person's identity after death.
2. Learning how to evaluate and properly act on the visual perceptions of a first impression is a skill. In his widely read book *Blink,* Malcolm Gladwell (2005) argues that first impressions are often correct and that the layers of cultural

rationalization we apply to people and situations to try to reverse first impressions can lead us to make major, potentially costly, and dangerous errors. Concerning the meaning of clothing, Harold Koda (2001) provides an excellent and profusely illustrated review of the "artifice of apparel," presented by region of the body.

3. Susan Benson's exploration of the cultural significance of permanent body modification in industrialized countries is one of the most insightful yet written, emphasizing the significance of the mutability of human notions of personhood and identity in modern societies (Benson 2000).
4. Reconstructive facial surgery became increasingly important in Europe during the early twentieth century, spurred primarily by the needs of World War I veterans, hundreds of thousands of whom suffered severe facial injuries as a result of trench warfare (Kemp 2004). Reconstructive plastic surgery became even more essential in World War II, as airmen in particular suffered disfiguring injuries and burns when airplane fuel exploded. Among the most famous reconstructive surgeons of that era was Archibald McIndoe, a New Zealander who pioneered new techniques for treating badly burned faces and hands while working in England. McIndoe fully appreciated the importance of the face as a social passport in human society and emphasized the social reintegration of people who had undergone reconstructive surgery. For a recent biography of McIndoe, see Mayhew 2004.
5. The age of the earliest sewn clothing can be inferred only by indirect evidence of the durable tools used to manufacture it or the beads or shells possibly used to ornament it. These tools and elements are associated with the Upper Paleolithic period (dating from about forty thousand to ten thousand years ago) (Bar-Yosef 2002). The earliest reliably dated needles appear to derive from the remnant of a hut located on the shores of the Sea of Galilee in modern Israel (Nadel et al. 2004), while assemblages of bone and shell beads used to decorate clothing and the body have been reported from sites in Turkey and Lebanon, dating from between forty-one and forty-three thousand years ago (Kuhn et al. 2001).
6. In his well-illustrated book (1997), Karl Groning reviews and authoritatively discusses the pigments used in body painting in different cultures. Christo-

pher Henshilwood and colleagues describe the discovery of the earliest documented use of red ochre in probable body decoration, in Blombos Cave in South Africa (Henshilwood et al. 2002). For good general discussions of the history of body painting, see Groning 1997; Walter et al. 1999.

7. Groning 1997.
8. Groning 1997; Walter et al. 1999.
9. Groning 1997.
10. White lead continued to be widely used in Europe through the 1820s, when it was replaced by zinc oxide. We will likely never know the extent of neurological morbidity or mortality that resulted from this source of lead poisoning.
11. The development of the face during childhood and its significance for relative eye size in adults is discussed in several studies; see Brown and Perrett 1993; Campbell et al. 1999; Schmidt and Cohn 2001.
12. Adding red coloring to the lips is a widespread practice of considerable antiquity, known to have occurred in Middle Kingdom Egypt and classical Greece. Reddened lips were popular in Europe from medieval times onward but were not always socially sanctioned. Meg Cohen Ragas and Karen Kozlowski (1998) have written a lively and well-illustrated history of lipstick.
13. Claudia Benthien (2002) discusses at length how allusions to facial coloration have been used in literature (especially in the novels of Honoré de Balzac) to convey the emotional states and psychological qualities of literary characters to the reader. On the meteoric rise of cosmetics sales and production, see Ragas and Kozlowski 1998.
14. Fowler 2000. The significance of Ötzi's tattoos and the meaning of their patterns have been the subjects of considerable scholarly and popular debate. Some scholars believe that the tattoos were therapeutic and not decorative.
15. Bogucki 1999.
16. Kirch 1997.
17. Benson (2000) argues that the functions of tattoos may have been particularly important to individuals or populations who had been stripped of their normal "social envelope" (such as recent immigrants to a country or a neighborhood) and who desired an obvious means of self-protection and self-identification. Connor (2004) provides an interesting discussion of the in-

nocence of unmarked skin in contrast to the permanent stain of the tattooed surface.

18. Benson (2000) explores the historical roots of modern Western tattooing and other forms of body modification and cogently summarizes the place of these practices in modern culture.
19. Numerous ethnographic accounts describe the use of tattooing to mark signal events in life history. William Saville (1926, 59–61) provides a particularly detailed description. Jean-Chris Miller (2004) offers an expansive discussion about what motivates people today to decorate themselves with permanent body art.
20. Laser removal of tattoos is generally more successful for darker colors of ink than for light ones, but the procedure is not without risk. Some lighter colors of ink (including two widely used azo compounds) break down into potentially toxic decomposition products when exposed to high-intensity laser irradiation (Vasold et al. 2004).
21. Based on interviews conducted by June Anderson of the California Academy of Sciences in March 1998 with mehndi artists Lila Kent and Ravie Kattaura.
22. Gay and Whittington 2002; Klesse 1999.
23. Stelarc quoted in Benthien 2002, 222. In a 1995 interview conducted by Paolo Atzori and Kirk Woolford of the Academy of Media Arts in Cologne, Germany, Stelarc stated, "I think metaphysically, in the past, we've considered the skin as surface, as interface. The skin has been a boundary for the soul, for the self, and simultaneously, a beginning to the world. Once technology stretches and pierces the skin, the skin as a barrier is erased." The full text of this interview is available online at www.ctheory.net/articles.aspx?id=71.
24. Caliendo, Armstrong, and Roberts 2005.
25. Miller 2004.
26. Groning 1997.
27. Numerous scholars have researched the association of dark skin and inferior social position, especially as it was used to justify the establishment and maintenance of the slave trade; see Babb 1998; Oakes 1998; Iyengar 2005. Lorenz Oken's *Lehrbuch der Naturphilosophie* (1811) and similar treatises on the cultural valuation of different colors are discussed cogently by Benthien

(2002). The imagery surrounding skin color and the awareness that skin has the character of a permanently worn garment with its own identity continue to infuse and inspire literature throughout the world; Benthien explores these topics in detail (2002, esp. chap. 8).

28. On the preference for light skin among dark-skinned populations in Africa and Melanesia, see Ardener 1954. Peter Frost of the Université Laval in Quebec discusses this topic in some detail; see Frost 2005, as well as his useful Web site, available at http://pages.globetrotter.net/peter_frost61z/. On the roots and ramifications of colorism in the United States, see Herring, Keith, and Horton 2004. Concerning the connection of light skin with infancy and femaleness, see Frost 2005; Frost 1988. Two excellent books discuss whiteness as a constructed social concept and the social desirability that has come to be associated with light skin color; see Babb 1998; Iyengar 2005.
29. Taylor 2002, 2003.
30. For lightly pigmented people, routine exposure to UVR and resultant damage to DNA in the skin carries an elevated risk of skin cancer (Cleaver and Crowley 2002; Sinni-McKeehen 1995; Christenson et al. 2005). As chapter 4 discussed, the connection between UVR exposure and the breakdown of folate has been documented in the laboratory; folate deficiency can develop in humans during the course of prolonged exposure to UVR (Branda and Eaton 1978; Jablonski and Chaplin 2000). This is likely the mechanism that produced neural tube defects found in fetuses carried by young women who frequented tanning parlors, as documented by Pablo Lapunzina (1996). The American Association of Dermatologists publishes a straightforward brochure on the risks of natural and indoor tanning; it is available online at www.aad.org/public/Publications/pamphlets/DarkerSideTanning.htm.
31. Monfrecola and Prizio 2001.
32. Brown 2001; Randle 1997; Monfrecola and Prizio 2001.
33. Anderson 1994.
34. "Cosmetic Enhancement Statistics at a Glance," *New Beauty*, Summer 2005, 25.
35. See Orlan's own Web site at www.orlan.net. The Web site of the Digitized

Bodies Project also provides a concise and insightful summary of Orlan's life and work to date; see www.digibodies.org/online/orlan.htm.

Chapter 11. Future Skin

1. For a summary of some of the problems involved in modern skin culture methods and the care of serious burn patients, see Dennis 2005. The technique that utilizes a person's own skin as skin grafts is now evolving into methods of using the person's skin to provide cells to "farm" new skin. Keratinocytes harvested from a stamp-sized piece of skin can be grown quickly in culture and then sprayed onto skin using a syringe fitted with a spray nozzle. "CellSpray," as it is called, can be used alone if the burn has not damaged large areas of the dermis or in conjunction with skin grafts if it has (Wood 2003).
2. Wood 2003.
3. Mansbridge 1999.
4. Zenz et al. 2005.
5. Wadman 2005; Martin and Parkhurst 2004.
6. This technology has been developed mostly by the VeriChip Corporation of Delray Beach, Florida. Their VeriChip RFID (radio frequency identification) microchip is small, about the size of a grain of rice, and is the only implantable chip currently approved for use in humans in the United States. Information on this technology is available through the company's Web site at www.verimedinfo.com/technology.html.
7. Implanting microchips that contain essential social and medical information has been promoted in some quarters as the safest, most convenient, and most reliable mode of personal identification. Many human rights organizations have challenged this idea, asserting that use of RFID chips violates individual privacy, that chips could be implanted against a person's will, and that the chips could be too easily reprogrammed to contain incorrect or deliberately incriminating information.
8. At the Baja Beach Club in Barcelona, Spain, customers buy an RFID microchip "debit card" when they arrive, have it implanted under their skin in a quick and nearly painless procedure, and party on until their funds are ex-

hausted. Party animals can add value to their chips at any time. The sophistication of such a system relies primarily on the specifications of the database to which it is attached; the RFID technology itself is relatively simple. See www.verichipcorp.com/content/solutions/verichip.

9. Sanchez-Vives and Slater 2005.
10. Investigators at the Interaction and Entertainment Research Centre at Nanyang Technical University in Singapore, using poultry as research models, have developed a "hug suit" that makes it possible to "touch" a distant chicken by stroking a model of the bird that is equipped with sensors. The sensors transmit impulses via the Internet to a jacket worn by the chicken that then vibrates, simulating touch. A short report on this innovation is available online at www2.ntu.edu.sg/ClassAct/Dec05/Research/1.htm.
11. Someya et al. 2005; Someya and Sakurai 2003; Someya et al. 2004; Cheung and Lumelsky 1992.
12. Manzotti and Tagliasco 2001.

REFERENCES

Abramson, Paul R., and Eldridge H. Pearsall. 1983. "Pectoral Changes during the Sexual Response Cycle: Thermographic Analysis." *Archives of Sexual Behavior* 12 (4): 357–368.

Adcock, Gregory J., Elizabeth S. Dennis, Simon Easteal, Gavin A. Huttley, Lars S. Jermiin, W. James Peacock, and Alan Thorne. 2001. "Mitochondrial DNA Sequences in Ancient Australians: Implications for Modern Human Origins." *Proceedings of the National Academy of Sciences U.S.A.* 98 (2): 537–542.

Alaluf, Simon, Derek Atkins, Karen Barrett, Margaret Blount, Nik Carter, and Alan Heath. 2002. "Ethnic Variation in Melanin Content and Composition in Photoexposed and Photoprotected Human Skin." *Pigment Cell Research* 15 (2): 112–118.

Altmann, Jeanne, Stuart A. Altmann, Glenn Hausfater, and Sue Ann McCuskey. 1977. "Life History of Yellow Baboons: Physical Development, Reproductive Parameters, and Infant Mortality." *Primates* 18 (2): 315–330.

Anderson, Mark M. 1994. *Kafka's Clothes: Ornament and Asceticism in the Habsburg Fin de Siècle.* Oxford: Oxford University Press.

Aoki, Kenichi. 2002. "Sexual Selection as a Cause of Human Skin Colour Variation: Darwin's Hypothesis Revisited." *Annals of Human Biology* 29 (6): 589–608.

Ardener, Edwin W. 1954. "Some Ibo Attitudes to Skin Pigmentation." *Man* 54 (101): 71–73.

Ardrey, Robert. 1961. *African Genesis: A Personal Investigation into the Animal Origins and Nature of Man*. New York: Simon and Schuster.

Aurelli, Filippo, and Frans B. M. de Waal, eds. 2000. *Natural Conflict Resolution*. Berkeley: University of California Press.

Autumn, Kellar, Metin Sitti, Yiching A. Liang, Anne M. Peattie, Wendy R. Hansen, Simon Sponberg, Thomas W. Kenny, Ronald Fearing, Jacob N. Israelachvili, and Robert J. Full. 2002. "Evidence for van der Waals Adhesion in Gecko Setae." *Proceedings of the National Academy of Sciences U.S.A.* 99 (19): 12252–12256.

Babb, Valerie. 1998. *Whiteness Visible: The Meaning of Whiteness in American Literature and Culture*. New York: New York University Press.

Baker, Paul T. 1958. "The Biological Adaptation of Man to Hot Deserts." *American Naturalist* 92 (867): 337–357.

Barber, Elizabeth. 2002. "Fashioned from Fiber." In *Along the Silk Road,* edited by E. Ten Grotenhuis, chap. 3. Washington, D.C.: Sackler Gallery, Smithsonian Institution.

Barker, Diane, Kathleen Dixon, Estela E. Medrano, Douglas Smalara, Sungbin Im, David Mitchell, George Babcock, and Zalfa A. Abdel-Malek. 1995. "Comparison of the Responses of Human Melanocytes with Different Melanin Contents to Ultraviolet B Irradiation." *Cancer Research* 55 (18): 4041–4046.

Barsh, Gregory. 1996. "The Genetics of Pigmentation: From Fancy Genes to Complex Traits." *Trends in Genetics* 12 (8): 299–305.

Bar-Yosef, Ofer. 2002. "The Upper Paleolithic Revolution." *Annual Review of Anthropology* 31:363–393.

Ben-Shakhar, Gershon, Maya Bar-Hillel, and Mordechai Kremnitzer. 2002. "Trial by Polygraph: Reconsidering the Use of the Guilty Knowledge Technique in Court." *Law and Human Behavior* 26 (5): 527–541.

Benson, Susan. 2000. "Inscriptions of the Self: Reflections on Tattooing and Piercing in Contemporary Euro-America." In *Written on the Body: The Tattoo in European and American History*, edited by Jane Caplan. London: Reaktion Books.

Benthien, Claudia 2002. *Skin: On the Cultural Border between Self and the World.* Translated by Thomas Dunlap. New York: Columbia University Press.

Biasutti, Renato. 1959. *Le razze e i popoli della terra: Razze, popoli, e culture.* 4 vols. Vol. 1, *Le razze e i popoli della terra.* Torino: Unione Tipografico-Editrice Torinese.

Blum, Harold F. 1961. "Does the Melanin Pigment of Human Skin Have Adaptive Value?" *Quarterly Review of Biology* 36:50–63.

Bögels, Susan M., and Caroline T. Lamers. 2002. "The Causal Role of Self-Awareness in Blushing-Anxious, Socially-Anxious, and Social Phobics Individuals." *Behaviour Research and Therapy* 40 (12): 1367–1384.

Bogucki, Peter 1999. *The Origins of Human Society.* Malden, Mass.: Blackwell.

Bower, Carol, and Fiona J. Stanley. 1989. "Dietary Folate as a Risk Factor for Neural-Tube Defects: Evidence from a Case-Control Study in Western Australia." *Medical Journal of Australia* 150 (11): 613–619.

Brace, C. Loring. 1963. "Structural Reduction in Evolution." *American Naturalist* 97 (892): 39–49.

Bramble, Dennis M., and Daniel E. Lieberman. 2004. "Endurance Running and the Evolution of *Homo*." *Nature* 432 (7015): 345–352.

Branda, Richard F., and John W. Eaton. 1978. "Skin Color and Nutrient Photolysis: An Evolutionary Hypothesis." *Science* 201 (4356): 625–626.

Brothwell, Don R. 1987. *The Bog Man and the Archaeology of People.* Cambridge, Mass.: Harvard University Press.

Brown, David A. 2001. "Skin Pigmentation Enhancers." In *Sun Protection in Man*, edited by Paolo U. Giacomoni. Amsterdam: Elsevier.

Brown, Elizabeth H., and David I. Perrett. 1993. "What Gives a Face Its Gender?" *Perception* 22 (7): 829–840.

Burgess, Ian F. 2004. "Human Lice and Their Control." *Annual Review of Entomology* 49:457–481.

Butts, Janie B. 2001. "Outcomes of Comfort Touch in Institutionalized Elderly Female Residents." *Geriatric Nursing* 22 (4): 180–184.

Cabanac, Michel, and Michael Caputa. 1979. "Natural Selective Cooling of the

Human Brain: Evidence of Its Occurrence and Magnitude." *Journal of Physiology* 286 (1): 255–264.

Cabanac, Michel, and B. Massonnet. 1977. "Thermoregulatory Responses as a Function of Core Temperature in Humans." *Journal of Physiology* 265 (1): 587–596.

Caldwell, Martyn M., Lars O. Björn, Janet F. Bornman, Stephan D. Flint, G. Kulandaivelu, Alan H. Teramura, and Manfred Tevini. 1998. "Effects of Increased Solar Ultraviolet Radiation on Terrestrial Ecosystems." *Journal of Photochemistry and Photobiology B: Biology* 46 (1–3): 40–52.

Caliendo, Carol, Myrna L. Armstrong, and Alden E. Roberts. 2005. "Self-Reported Characteristics of Women and Men with Intimate Body Piercings." *Journal of Advanced Nursing* 49 (5): 474–484.

Campbell, Ruth, Philip J. Benson, Simon B. Wallace, Suzanne Doesbergh, and Michael Coleman. 1999. "More about Brows: How Poses That Change Brow Position Affect Perceptions of Gender." *Perception* 28 (4): 489–504.

Caputa, Michael, and Michel Cabanac. 1988. "Precedence of Head Homoeothermia over Trunk Homoeothermia in Dehydrated Men." *European Journal of Applied Physiology* 57 (5): 611–613.

Carpenter, Peter W., Christopher Davies, and Anthony D. Lucey. 2000. "Hydrodynamics and Compliant Walls: Does the Dolphin Have a Secret?" *Current Science* 79 (6): 758–765.

Carrier, David R. 1984. "The Energetic Paradox of Human Running and Hominid Evolution." *Current Anthropology* 25 (4): 483–495.

Chaplin, George. 2001. "The Geographic Distribution of Environmental Factors Influencing Human Skin Colouration." MSc thesis, Manchester Metropolitan University.

———. 2004. "Geographic Distribution of Environmental Factors Influencing Human Skin Coloration." *American Journal of Physical Anthropology* 125 (3): 292–302.

Chaplin, George, and Nina G. Jablonski. 1998. "Hemispheric Difference in Human Skin Color." *American Journal of Physical Anthropology* 107 (2): 221–223.

Chaplin, George, Nina G. Jablonski, and N. Timothy Cable. 1994. "Physiology, Thermoregulation, and Bipedalism." *Journal of Human Evolution* 27 (6): 497–510.

Cheung, Edward, and Vladimir Lumelsky. 1992. "Sensitive Skin System for Motion Control of Robot Arm Manipulators." *Robotics and Autonomous Systems* 10 (1): 9–32.

Chiappe, Luis M. 1995. "The First 85 Million Years of Avian Evolution." *Nature* 378 (6555): 349–355.

Chiappe, Luis M., Rodolfo A. Coria, Lowell Dingus, Frankie Jackson, Anusuya Chinsamy, and Marilyn Fox. 1998. "Sauropod Dinosaur Embryos from the Late Cretaceous of Patagonia." *Nature* 396 (6708): 258–261.

Chimpanzee Sequencing and Analysis Consortium. 2005. "Initial Sequence of the Chimpanzee Genome and Comparison with the Human Genome." *Nature* 437 (7055): 69–87.

Christenson, Leslie J., Theresa A. Borrowman, Celine M. Vachon, Megha M. Tollefson, Clark C. Otley, Amy L. Weaver, and Randall K. Roenigk. 2005. "Incidence of Basal Cell and Squamous Cell Carcinomas in a Population Younger Than 40 Years." *Journal of the American Medical Association* 294 (6): 681–690.

Chu, David H., Anne R. Haake, Karen Holbrook, and Cynthia A. Loomis. 2003. "The Structure and Development of Skin." In *Fitzpatrick's Dermatology in General Medicine,* edited by Irwin M. Freedberg, Arthur Z. Eisen, Klaus Wolff, K. Frank Austen, Lowell A. Goldsmith, and Stephen I. Katz. 6th ed. New York: McGraw-Hill.

Chuang, Tsu-Yi, George T. Reizner, David J. Elpern, Jenny L. Stone, and Evan R. Farmer. 1995. "Nonmelanoma Skin Cancer in Japanese Ethnic Hawaiians in Kauai, Hawaii: An Incidence Report." *Journal of the American Academy of Dermatology* 33 (3): 422–426.

Chuong, Cheng-Ming, Ping Wu, Fu-Cheng Zhang, Xing Xu, Minke Yu, Randall B. Widelitz, Ting-Xin Jiang, and Lianhai Hou. 2003. "Adaptation to the Sky: Defining the Feather with Integument Fossils from Mesozoic China and Experimental Evidence from Molecular Laboratories." *Journal of Experimental Zoology, Part B, Molecular and Developmental Evolution* 298 (1): 42–56.

Clarke, Barry T. 1997. "The Natural History of Amphibian Skin Secretions, Their Normal Functioning and Potential Medical Applications." *Biological Reviews of the Cambridge Philosophical Society* 72 (3): 365–379.

Cleaver, James E., and Eileen Crowley. 2002. "UV Damage, DNA Repair, and Skin Carcinogenesis." *Frontiers in Bioscience* 7:1024–1043.

Collet, Christian, Evelyne Vernet-Maury, Georges Delhomme, and André Dittmar. 1997. "Autonomic Nervous System Response Patterns Specificity to Basic Emotions." *Journal of the Autonomic Nervous System* 62 (1–2): 45–57.

Connor, Steven. 2004. *The Book of Skin.* Ithaca, N.Y.: Cornell University Press.

Cornish, Daryl A., Vusi Maluleke, and Thulani Mhlanga. 2000. "An Investigation into a Possible Relationship between Vitamin D, Parathyroid Hormone, Calcium, and Magnesium in a Normally Pigmented and an Albino Rural Black Population in the Northern Province of South Africa." *BioFactors* 11 (1–2): 35–38.

Cosentino, M. James, Ruth E. Pakyz, and Josef Fried. 1990. "Pyrimethamine: An Approach to the Development of a Male Contraceptive." *Proceedings of the National Academy of Sciences U.S.A.* 87 (4): 1431–1435.

Cowles, Raymond B. 1959. "Some Ecological Factors Bearing on the Origin and Evolution of Pigment in the Human Skin." *American Naturalist* 93 (872): 283–293.

Daniels, Farrington. 1964. "Man and Radiant Energy: Solar Radiation." In *Handbook of Physiology, Section 4: Adaptation to the Environment,* edited by D. B. Dill, E. F. Adolph, and C. G. Wilber. Washington, D.C.: American Physiological Society.

de Gruijl, Frank R., and Henk J. van Kranen. 2001. "UV Radiation, Mutations, and Oncogenic Pathways in Skin Cancer." In *Sun Protection in Man,* edited by Paolo U. Giacomoni. Amsterdam: Elsevier.

Dennis, Carina. 2005. "Spray-On Skin: Hard Graft." *Nature* 436 (7048): 166–167.

de Waal, Frans B. M. 1990. *Peacemaking among Primates.* Cambridge, Mass.: Harvard University Press.

———. 1993. "Reconciliation among Primates: A Review of Empirical Evidence and Unresolved Issues." In *Primate Social Conflict,* edited by William A. Mason and Sally P. Mendoza. New York: SUNY Press.

de Waal, Frans B. M., and Angeline van Roosmalen. 1979. "Reconciliation and Consolation among Chimpanzees." *Behavioral Ecology and Sociobiology* 5 (1): 55–66.

Diamond, Jared. 2005. *Collapse: How Societies Choose to Fail or Succeed.* New York: Viking Press.

Diegelmann, Robert F., and Melissa C. Evans. 2004. “Wound Healing: An Overview of Acute, Fibrotic, and Delayed Healing.” *Frontiers in Bioscience* 9:283–289.

Di Folco, Philippe. 2004. *Skin Art*. Paris: Fitway Publishing.

Dillard, James N., and Sharon Knapp. 2005. “Complementary and Alternative Pain Therapy in the Emergency Department.” *Emergency Medicine Clinics of North America* 23 (2): 529–549.

Ding, Mei, Wei-Meng Woo, and Andrew D. Chisholm. 2004. “The Cytoskeleton and Epidermal Morphogenesis in *C. elegans*.” *Experimental Cell Research* 301 (1): 84–90.

Domb, Leah G., and Mark Pagel. 2001. “Sexual Swellings Advertise Female Quality in Wild Baboons.” *Nature* 410 (6825): 204–206.

Dominy, Nathaniel J. 2004. “Fruits, Fingers, and Fermentation: The Sensory Cues Available to Foraging Primates.” *Integrative and Comparative Biology* 44 (4): 295–303.

Dominy, Nathaniel J., and Peter W. Lucas. 2001. “Ecological Importance of Trichromatic Vision to Primates.” *Nature* 410 (6826): 363–366.

Drummond, Peter D. 1999. “Facial Flushing during Provocation in Women.” *Psychophysiology* 36 (3): 325–332.

Drummond, Peter D., and James W. Lance. 1987. “Facial Flushing and Sweating Mediated by the Sympathetic Nervous System.” *Brain* 110 (3): 793–803.

Drummond, Peter D., and Saw Han Quah. 2001. “The Effect of Expressing Anger on Cardiovascular Reactivity and Facial Blood Flow in Chinese and Caucasians.” *Psychophysiology* 38 (2): 190–196.

Dunbar, Robin I. M. 2001. “What’s in a Baboon’s Behind?” *Nature* 410 (6825): 158.

Ekman, Paul. 1998. “Introduction to the Third Edition.” In *The Expression of the Emotions in Man and Animals,* by Charles Darwin. New York: Oxford University Press.

———. 2003. *Emotions Revealed: Recognizing Faces and Feelings to Improve Communication and Emotional Life*. New York: Time Books.

Elias, Peter M., Kenneth R. Feingold, and Joachim W. Fluhr. 2003. “Skin as an Organ of Protection.” In *Fitzpatrick’s Dermatology in General Medicine,* edited by Irwin M. Freedberg, Arthur Z. Eisen, Klaus Wolff, K. Frank Austen, Lowell A. Goldsmith, and Stephen I. Katz. 6th ed. New York: McGraw-Hill.

Eliot, T. S. 1920. *Poems*. New York: Knopf.

Erb, Peter, Jingmin Ji, Marion Wernli, Erwin Kump, Andrea Glaser, and Stanislaw A. Buchner. 2005. "Role of Apoptosis in Basal Cell and Squamous Cell Carcinoma Formation." *Immunology Letters* 100 (1): 68–72.

Falk, Dean. 1990. "Brain Evolution in *Homo:* The 'Radiator' Theory." *Behavioral and Brain Sciences* 13:333–381.

Field, Tiffany. 2001. *Touch*. Cambridge, Mass.: MIT Press.

———. 2002. "Violence and Touch Deprivation in Adolescents." *Adolescence* 37 (148): 735–749.

Fisher, Gary J., Sewon Kang, James Varani, Zsuzsanna Bata-Csorgo, Wen Yinsheng, Subhash Datta, and John J. Voorhees. 2002. "Mechanisms of Photoaging and Chronological Skin Aging." *Archives of Dermatology* 138 (11): 1462–1470.

Fitzpatrick, Thomas B., Makoto Seiji, and A. David McGugan. 1961. "Melanin Pigmentation." *New England Journal of Medicine* 265 (7): 328–332.

Fitzpatrick, Thomas R., and Jean-Paul Ortonne. 2003. "Normal Skin Color and General Considerations of Pigmentary Disorders." In *Fitzpatrick's Dermatology in General Medicine,* edited by Irwin M. Freedberg, Arthur Z. Eisen, Klaus Wolff, K. Frank Austen, Lowell A. Goldsmith, and Stephen I. Katz. 6th ed. New York: McGraw-Hill.

Fleming, Angeleen, and Andrew J. Copp. 1998. "Embryonic Folate Metabolism and Mouse Neural Tube Defects." *Science* 280 (5372): 2107–2109.

Folk, G. Edgar, Jr., and Holmes A. Semken Jr. 1991. "The Evolution of Sweat Glands." *International Journal of Biometeorology* 35 (3): 180–186.

Fowler, Brenda. 2000. *Iceman: Uncovering the Life and Times of a Prehistoric Man Found in an Alpine Glacier.* Chicago: University of Chicago Press.

Frisancho, A. Roberto. 1995. *Human Adaptation and Accommodation*. Rev. ed. Ann Arbor: University of Michigan Press.

Frost, Peter. 1988. "Human Skin Color: A Possible Relationship between Its Sexual Dimorphism and Its Social Perception." *Perspectives in Biology and Medicine* 32 (1): 38–59.

———. 2005. *Fair Women, Dark Men: The Forgotten Roots of Color Prejudice*. N.p.: Cybereditions.

Garland, Cedric F., Frank C. Garland, and Edward D. Gorham. 2003. "Epidemiologic Evidence for Different Roles of Ultraviolet A and B Radiation in Melanoma Mortality Rates." *Annals of Epidemiology* 13 (6): 395–404.

Garland, Cedric F., Frank C. Garland, Edward D. Gorham, Martin Lipkin, Harold Newmark, Sharif B. Mohr, and Michael F. Holick. 2006. "The Role of Vitamin D in Cancer Prevention." *American Journal of Public Health* 96 (2): 252–261.

Gay, Kathlyn, and Christine Whittington. 2002. *Body Marks: Tattooing, Piercing, and Scarification*. Brookfield, Conn.: Millbrook Press.

Gessner, Bradford D., Elizabeth deSchweinitz, Kenneth M. Petersen, and Christopher Lewandowski. 1997. "Nutritional Rickets among Breast-Fed Black and Alaska Native Children." *Alaska Medicine* 39 (3): 72–87.

Gharaee-Kermani, Mehrnaz, and Sem H. Phan. 2001. "Role of Cytokines and Cytokine Therapy in Wound Healing and Fibrotic Diseases." *Current Pharmaceutical Design* 7 (11): 1083–1103.

Givler, Robert C. 1920. *Psychology: The Science of Human Behavior.* New York: Harper.

Gladwell, Malcolm. 2005. *Blink: The Power of Thinking without Thinking.* New York: Little, Brown.

Gleeson, Madeline, and Fiona Timmins. 2004. "The Use of Touch to Enhance Nursing Care of Older Person in Long-Term Mental Health Care Facilities." *Journal of Psychiatric and Mental Health Nursing* 11 (5): 541–545.

Goldsmith, Lowell A. 2003. "Biology of Eccrine and Apocrine Sweat Glands." In *Fitzpatrick's Dermatology in General Medicine,* edited by Irwin M. Freedberg, Arthur Z. Eisen, Klaus Wolff, K. Frank Austen, Lowell A. Goldsmith, and Stephen I. Katz. 6th ed. New York: McGraw-Hill.

Goodall, Jane. 1986. *The Chimpanzees of Gombe: Patterns of Behavior.* Cambridge, Mass.: Belknap Press of Harvard University Press.

Grandin, Temple. 1996. *Thinking in Pictures and Other Reports from My Life with Autism.* New York: Vintage Books.

Grant, William B. 2003. "Ecologic Studies of Solar UV-B Radiation and Cancer Mortality Rates." *Recent Results in Cancer Research* 164:371–377.

Gravlee, Clarence C., William W. Dressler, and H. Russell Bernard. 2005. "Skin Color, Social Classification, and Blood Pressure in Southeastern Puerto Rico." *American Journal of Public Health* 95 (12): 2191–2197.

Grevelink, Suzanne Virnelli, and John Butler Mulliken. 2003. "Vascular Anomalies and Tumors of Skin and Subcutaneous Tissues." In *Fitzpatrick's Dermatology in General Medicine,* edited by Irwin M. Freedberg, Arthur Z. Eisen, Klaus Wolff, K. Frank Austen, Lowell A. Goldsmith, and Stephen I. Katz. 6th ed. New York: McGraw-Hill.

Groning, Karl. 1997. *Decorated Skin: A World Survey of Body Art.* London: Thames and Hudson.

Halaban, Ruth, Daniel N. Hebert, and David E. Fisher. 2003. "Biology of Melanocytes." In *Fitzpatrick's Dermatology in General Medicine,* edited by Irwin M. Freedberg, Arthur Z. Eisen, Klaus Wolff, K. Frank Austen, Lowell A. Goldsmith, and Stephen I. Katz. 6th ed. New York: McGraw-Hill.

Hardy, Alister. 1960. "Was Man More Aquatic in the Past?" *New Scientist* 7:642–645.

Harlow, Harry F., Margaret K. Harlow, Robert O. Dodsworth, and G. L. Arling. 1966. "Maternal Behavior of Rhesus Monkeys Deprived of Mothering and Peer Associations in Infancy." *Proceedings of the American Philosophical Society* 110 (1): 58–66.

Harlow, Harry F., and Robert R. Zimmerman. 1958. "The Development of Affectional Responses in Infant Monkeys." *Proceedings of the American Philosophical Society* 102 (5): 501–509.

Hartl, Daniel L. 2004. "The Origin of Malaria: Mixed Messages from Genetic Diversity." *Nature Reviews: Microbiology* 2 (1): 15–22.

Haworth, James C., and Louise A. Dilling. 1986. "Vitamin-D-Deficient Rickets in Manitoba, 1972–84." *Canadian Medical Association Journal* 134 (3): 237–241.

Healy, Eugene, Siobhan A. Jordan, Peter S. Budd, Ruth Suffolk, Jonathan L. Rees, and Ian J. Jackson. 2001. "Functional Variation of *MC1R* Alleles from Red-Haired Individuals." *Human Molecular Genetics* 10 (21): 2397–2402.

Henshilwood, Christopher S., Francesco d'Errico, Royden Yates, Zenobia Jacobs, Chantal Tribolo, Geoff A. T. Duller, Norbert Mercier, Judith C. Sealy, Helene Valladas, Ian Watts, and Ann G. Wintle. 2002. "Emergence of Modern Human Behavior: Middle Stone Age Engravings from South Africa." *Science* 295 (5558): 1278–1280.

Herman, Jay R., and Edward A. Celarier. 1996. "TOMS Version 7 UV-Erythemal Exposure: 1978–1993." CD-ROM. Edited by NASA. Goddard Space Flight Center.

Herring, Cedric, Verna M. Keith, and Hayward Derrick Horton, eds. 2004. *Skin/Deep: How Race and Complexion Matter in the "Color-Blind" Era.* Urbana: Institute for Research on Race and Public Policy and University of Illinois Press.

Hirakawa, Kazutaka, Hiroyuki Suzuki, Shinji Oikawa, and Shosuke Kawanishi. 2002. "Sequence-Specific DNA Damage Induced by Ultraviolet A–Irradiated Folic Acid via Its Photolysis Product." *Archives of Biochemistry and Biophysics* 410 (2): 261–268.

Hitchcock, R. Timothy. 2001. *Ultraviolet Radiation.* 2nd ed. Nonionizing Radiation Guide Series. Fairfax, Va.: American Industrial Hygiene Association.

Hodgkin, P., G. H. Kay, P. M. Hine, G. A. Lumb, and S. W. Stanbury. 1973. "Vitamin-D Deficiency in Asians at Home and in Britain." *The Lancet* 2 (7822): 167–172.

Holick, Michael F. 1987. "Photosynthesis of Vitamin D in the Skin: Effect of Environmental and Life-Style Variables." *Federation Proceedings* 46 (5): 1876–1882.

———. 1995. "Environmental Factors That Influence the Cutaneous Production of Vitamin D." *American Journal of Clinical Nutrition* 61 (3 suppl.): 638S–645S.

———. 1997. "Photobiology of Vitamin D." In *Vitamin D,* edited by David Feldman, Francis H. Glorieux, and J. Wesley Pike. San Diego: Academic Press.

———. 2001. "A Perspective on the Beneficial Effects of Moderate Exposure to Sunlight: Bone Health, Cancer Prevention, Mental Health, and Well Being." In *Sun Protection in Man,* edited by Paolo U. Giacomoni. Amsterdam: Elsevier.

———. 2003. "Evolution and Function of Vitamin D." *Recent Results in Cancer Research* 164:3–28.

———. 2004. "Vitamin D: Importance in the Prevention of Cancers, Type 1 Diabetes, Heart Disease, and Osteoporosis." *American Journal of Clinical Nutrition* 79 (3): 362–371.

Holick, Michael F., Julia A. MacLaughlin, and S. H. Doppelt. 1981. "Regulation of Cutaneous Previtamin D_3 Photosynthesis in Man: Skin Pigment Is Not an Essential Regulator." *Science* 211 (4482): 590–593.

Homsi, Jade, Mohammed Kashani-Sabet, Jane L. Messina, and Adil Daud. 2005. "Cutaneous Melanoma: Prognostic Factors." *Cancer Control* 12 (4): 223–229.

Horiuchi, Shiro. 2005. "Affiliative Relations among Male Japanese Macaques *(Macaca fuscata yakui)* within and outside a Troop on Yakushima Island." *Primates* 46 (3): 191–197.

Ikoma, Akihiko, Roman Rukwied, Sonja Ständer, Martin Steinhoff, Yoshiki Miyachi, and Martin Schmelz. 2003. "Neurophysiology of Pruritus: Interaction of Itch and Pain." *Archives of Dermatology* 139 (11): 1475–1478.

Ito, Shosuke. 2003. "A Chemist's View of Melanogenesis." *Pigment Cell Research* 16 (3): 230–236.

Iyengar, Sujata. 2005. *Shades of Difference: Mythologies of Skin Color in Early Modern England.* Philadelphia: University of Pennsylvania Press.

Jablonski, Nina G. 2004. "The Evolution of Human Skin and Skin Color." *Annual Review of Anthropology* 33:585–623.

Jablonski, Nina G., and George Chaplin. 1993. "Origin of Habitual Terrestrial Bipedalism in the Ancestor of the Hominidae." *Journal of Human Evolution* 24 (4): 259–280.

———. 2000. "The Evolution of Skin Coloration." *Journal of Human Evolution* 39 (1): 57–106.

Jablonski, Nina G., George Chaplin, and Kenneth J. McNamara. 2002. "Natural Selection and the Evolution of Hominid Patterns of Growth and Development." In *Human Evolution through Developmental Change,* edited by Nancy Minugh-Purvis and Kenneth J. McNamara. Baltimore: Johns Hopkins University Press.

Jerison, Harry J. 1978. "Allometry and Encephalization." In *Recent Advances in Primatology,* vol. 3, *Evolution,* edited by D. J. Chivers and K. A. Joysey. London: Academic Press.

———. 1997. "Evolution of Prefrontal Cortex." In *Development of the Prefrontal Cortex: Evolution, Neurobiology, and Behavior,* edited by Norman A. Krasnegor, G. Reid Lyon, and Patricia S. Rakic. Baltimore: Paul H. Brookes.

John, Premila R., Kateryna Makova, Wen-Hsiung Li, Trefor Jenkins, and Michele Ramsay. 2003. "DNA Polymorphism and Selection at the Melanocortin-1 Receptor Gene in Normally Pigmented Southern African Individuals." *Annals of the New York Academy of Sciences* 994:299–306.

Johnson, Francis S., Tsan Mo, and Alex E. S. Green. 1976. "Average Latitudinal

Variation in Ultraviolet Radiation at the Earth's Surface." *Photochemistry and Photobiology* 23:179–188.

Kaidbey, Kays H., Patricia Poh Agin, Robert M. Sayre, and Albert M. Kligman. 1979. "Photoprotection by Melanin: A Comparison of Black and Caucasian Skin." *American Academy of Dermatology* 1 (3): 249–260.

Kalkwarf, Heidi J., and Bonny L. Specker. 2002. "Bone Mineral Changes during Pregnancy and Lactation." *Endocrine* 17 (1): 49–53.

Kappes, Ulrike P., Dan Luo, Marisa Potter, Karl Schulmeister, and Thomas M. Runger. 2006. "Short- and Long-Wave UV Light (UVB and UVA) Induce Similar Mutations in Human Skin Cells." *Journal of Investigative Dermatology* 126 (3): 667–675.

Katsarou-Katsari, Alexandra, A. Filippou, and Theoharis C. Theoharides. 1999. "Effect of Stress and Other Psychological Factors on the Pathophysiology and Treatment of Dermatoses." *International Journal of Immunopathology and Pharmacology* 12 (1): 7–11.

Kemp, Sandra. 2004. *Future Face: Image, Identity, Innovation.* London: Profile Books.

Kesavan, Vellappan, Madan S. Pote, Vipen Batra, and Gomathy Viswanathan. 2003. "Increased Folate Catabolism Following Total Body Y-Irradiation in Mice." *Journal of Radiation Research* 44 (2): 141–144.

Ketchum, L. D., I. K. Cohen, and F. W. Masters. 1974. "Hypertrophic Scars and Keloidal Scars." *Plastic and Reconstructive Surgery* 53 (2): 140–154.

Khaitovich, Philipp, Ines Hellmann, Wolfgang Enard, Katja Nowick, Marcus Leinweber, Henriette Franz, Gunter Weiss, Michael Lachmann, and Svante Pääbo. 2005. "Parallel Patterns of Evolution in the Genomes and Transcriptomes of Humans and Chimpanzees." *Science* 309 (5742): 1850–1854.

Kirch, Patrick V. 1997. *The Lapita Peoples: Ancestors of the Oceanic World.* Malden, Mass.: Blackwell.

Klein, Richard G., Graham Avery, Kathryn Cruz-Uribe, David Halkett, John E. Parkington, Teresa Steele, Thomas P. Volman, and Royden Yates. 2004. "The Ysterfontein 1 Middle Stone Age Site, South Africa, and Early Human Exploitation of Coastal Resources." 10.1073/pnas.0400528101. *Proceedings of the National Academy of Sciences U.S.A.* 101 (16): 5708–5715.

Klesse, Christian. 1999. "'Modern Primitivism': Non-Mainstream Body Modification and Racialized Representation." *Body and Society* 5 (2–3): 15–38.
Kligman, Albert M. 2004. "A Personal Critique on the State of Knowledge of Rosacea." *Dermatology* 208 (3): 191–197.
Knip, Agatha S. 1977. "Ethnic Studies on Sweat Gland Counts." In *Physiological Variation and Its Genetic Basis,* edited by J. S. Weiner. London: Taylor and Francis.
Koda, Harold. 2001. *Extreme Beauty: The Body Transformed.* New York: Metropolitan Museum of Art.
Kollias, Nikiforos. 1995a. "The Physical Basis of Skin Color and Its Evaluation." *Clinics in Dermatology* 13 (4): 361–367.
———. 1995b. "The Spectroscopy of Human Melanin in Pigmentation." In *Melanin: Its Role in Human Photoprotection,* edited by Lisa Zeise, Miles R. Chedekel, and Thomas B. Fitzpatrick. Overland Park, KS: Valdenmar Publications.
Kollias, Nikiforos, Robert M. Sayre, Lisa Zeise, and Miles R. Chedekel. 1991. "New Trends in Photobiology: Photoprotection by Melanin." *Journal of Photochemistry and Photobiology B* 9 (2): 135–160.
Kovacs, Christopher S. 2005. "Calcium and Bone Metabolism during Pregnancy and Lactation." *Journal of Mammary Gland Biology and Neoplasia* 10 (2): 105–118.
Kuhn, Steven L., Mary C. Stiner, David S. Reese, and Erksin Güleç. 2001. "Ornaments of the Earliest Upper Paleolithic: New Insights from the Levant." *Proceedings of the National Academy of Sciences U.S.A.* 98 (13): 7641–7646.
Lamason, Rebecca L., Manzoor-Ali P. K. Mohideen, Jason R. Mest, Andrew C. Wong, Heather L. Norton, Michele C. Aros, Michael J. Jurynec, Xianyun Mao, Vanessa R. Humphreville, Jasper E. Humbert, Soniya Sinha, Jessica L. Moore, Pudur Jagadeeswaran, Wei Zhao, Gang Ning, Izabela Makalowska, Paul M. McKeigue, David O'Donnell, Rick Kittles, Esteban J. Parra, Nancy J. Mangini, David J. Grunwald, Mark D. Shriver, Victor A. Canfield, and Keith C. Cheng. 2005. "SLC24A5, a Putative Cation Exchanger, Affects Pigmentation in Zebrafish and Humans." 10.1126/science.1116238. *Science* 310 (5755): 1782–1786.
Lapunzina, Pablo. 1996. "Ultraviolet Light–Related Neural Tube Defects?" *American Journal of Medical Genetics Part B Neuropsychiatric Genetics* 67 (1): 106.
Leary, Mark R., Thomas W. Britt, William D. Cutlip II, and Janice L. Templeton. 1992. "Social Blushing." *Psychological Bulletin* 112 (3): 446–460.

Lee, Marjorie M. C., and Gabriel W. Lasker. 1959. "The Sun-Tanning Potential of Human Skin." *Human Biology* 31:252–260.

Lock-Andersen, Jørgen, N. Ditlev Knudstorp, and Hans Christian Wulf. 1998. "Facultative Skin Pigmentation in Caucasians: An Objective Biological Indicator of Lifetime Exposure to Ultraviolet Radiation?" *British Journal of Dermatology* 138 (5): 826–832.

Loomis, W. Farnsworth. 1967. "Skin-Pigment Regulation of Vitamin-D Biosynthesis in Man." *Science* 157 (3788): 501–506.

Luck, C. P., and P. G. Wright. 1964. "Aspects of the Anatomy and Physiology of the Skin of the Hippopotamus *(H. amphibius)*." *Quarterly Journal of Experimental Physiology and Cognate Medical Sciences* 49 (1): 1–14.

Lucock, Mark, Zoe Yates, Tracey Glanville, Robert Leeming, Nigel Simpson, and Ioannis Daskalakis. 2003. "A Critical Role for B-Vitamin Nutrition in Human Development and Evolutionary Biology." *Nutrition Research* 23 (11): 1463–1475.

Luis, J. R., Diane J. Rowold, M. Regueiro, B. Caeiro, Cengiz Cinnioğlu, Charles Roseman, Peter A. Underhill, L. Luca Cavalli-Sforza, and Rene J. Herrera. 2004. "The Levant versus the Horn of Africa: Evidence for Bidirectional Corridors of Human Migrations." *American Journal of Human Genetics* 74 (3): 532–544.

MacLaughlin, Julia A., R. R. Anderson, and Michael F. Holick. 1982. "Spectral Character of Sunlight Modulates Photosynthesis of Previtamin D_3 and Its Photoisomers in Human Skin." *Science* 216 (4549): 1001–1003.

Madronich, Sasha, Richard L. McKenzie, Lars O. Björn, and Martyn M. Caldwell. 1998. "Changes in Biologically Active Ultraviolet Radiation Reaching the Earth's Surface." *Journal of Photochemistry and Photobiology B: Biology* 46 (1–3): 5–19.

Mahoney, Sheila A. 1980. "Cost of Locomotion and Heat Balance during Rest and Running from 9 to 55 C in a Patas Monkey." *Journal of Applied Physiology* 49 (5): 789–800.

Mansbridge, Jonathan. 1999. "Tissue-Engineered Skin Substitutes." *Expert Opinion on Investigative Drugs* 8 (7): 957–962.

Manson, Joseph H., and Richard W. Wrangham. 1991. "Intergroup Aggression in Chimpanzees and Humans." *Current Anthropology* 32 (4): 369–391.

Manzotti, Riccardo, and Vincenzo Tagliasco. 2001. "On Building an Artificial Conscious Being." Paper presented at In Search of a Science of Consciousness conference. Skovde, Sweden.

Marks, Jonathan. 2003. *What It Means to Be 98% Chimpanzee: Apes, People, and Their Genes.* Berkeley: University of California Press.

Martin, Paul, and Susan M. Parkhurst. 2004. "Parallels between Tissue Repair and Embryo Morphogenesis." *Development* 131 (13): 3021–3034.

Mathur, U., S.L. Datta, and B.B. Mathur. 1977. "The Effect of Aminopterin-Induced Folic Acid Deficiency on Spermatogenesis." *Fertility and Sterility* 28 (12): 1356–1360.

Matsumoto-Oda, Akiko. 1998. "Injuries to the Sexual Skin of Female Chimpanzees at Mahale and Their Effect on Behaviour." *Folia Primatologica* 69 (6): 400–404.

Matsumura, Yashuhiro, and Honnavara N. Ananthawamy. 2004. "Toxic Effects of Ultraviolet Radiation on the Skin." *Toxicology and Applied Pharmacology* 195 (3): 298–308.

Mayhew, Emily. 2004. *The Reconstruction of Warriors: Archibald McIndoe, the Royal Air Force, and the Guinea Pig Club.* London: Greenhill Books.

Miller, Greg. 2005. "Bats Have a Feel for Flight." *Science* 310 (5752): 1260–1261.

Miller, Jean-Chris. 2004. *The Body Art Book: A Complete Illustrated Guide to Tattoos, Piercings, and Other Body Modifications.* New York: Berkley.

Moffatt, Michael E. K. 1995. "Current Status of Nutritional Deficiencies in Canadian Aboriginal People." *Canadian Journal of Physiology and Pharmacology* 73 (6): 754–758.

"The Molecular Biology of Wound Healing." 2004. *Public Library of Science: Biology* 2 (8): e278.

Monfrecola, Giuseppe, and Emilia Prizio. 2001. "Self Tanning." In *Sun Protection in Man,* edited by Paolo U. Giacomoni. Amsterdam: Elsevier.

Montagna, William. 1971. "Cutaneous Comparative Biology." *Archives of Dermatology* 104 (6): 577–591.

———. 1981. "The Consequences of Having a Naked Skin." *Birth Defects: Original Article Series* 17 (2): 1–7.

Montagu, Ashley. 1971. *Touching: The Human Significance of the Skin.* New York: Columbia University Press.

Montoya, Pedro, J. Javier Campos, and Rainer Schandry. 2005. "See Red? Turn Pale? Unveiling Emotions through Cardiovascular and Hemodynamic Changes." *Spanish Journal of Psychology* 8 (1): 79–85.

Morbeck, Mary Ellen, Adrienne L. Zihlman, and Alison Galloway. 1993. "Biographies Read in Bones: Biology and Life History of Gombe Chimpanzees." In *Proceedings of the 1992 ChimpanZoo Conference,* edited by V. Landau. Jane Goodall Institute.

Morgan, Elaine. 1982. *The Aquatic Ape.* London: Souvenir.

Murray, Frederick G. 1934. "Pigmentation, Sunlight, and Nutritional Disease." *American Anthropologist* 36 (3): 438–445.

Nadel, Dani, Ehud Weiss, Orit Simchoni, Alexander Tsatskin, Avinoam Danin, and Mordechai Kislev. 2004. "Stone Age Hut in Israel Yields World's Oldest Evidence of Bedding." *Proceedings of the National Academy of Sciences U.S.A.* 101 (17): 6821–6826.

Nelson, David A., and Sarah A. Nunneley. 1998. "Brain Temperature and Limits on Transcranial Cooling in Humans: Quantitative Modeling Results." *European Journal of Applied Physiology* 78 (4): 353–359.

Newton Bishop, Julia A., and D. Timothy Bishop. 2005. "The Genetics of Susceptibility to Cutaneous Melanoma." *Drugs of Today* 41 (3): 193–203.

Oakes, James. 1998. *The Ruling Race: A History of American Slaveholders.* New York: Norton.

Off, Morten Christian, Arnfinn Engeset Steindal, Alina Carmen Porojnicu, Asta Juzeniene, Alexander Vorobey, Anders Johnsson, and Johan Moan. 2005. "Ultraviolet Photodegradation of Folic Acid." *Journal of Photochemistry and Photobiology B: Biology* 80 (1): 47–55.

Olivier, Georges. 1960. *Pratique anthropologique.* Paris: Vigot Frères, Editeurs.

Ortonne, Jean-Paul. 2002. "Photoprotective Properties of Skin Melanin." *British Journal of Dermatology* 146 (suppl. 61): 7–10.

Padian, Kevin. 2001. "Cross-Testing Adaptive Hypotheses: Phylogenetic Analysis and the Origin of Bird Flight." *American Zoologist* 41 (3): 598–607.

Pagel, Mark, and Walter Bodmer. 2003. "A Naked Ape Would Have Fewer Parasites." *Proceedings of the Royal Society of London B* 270 (suppl.): S117–S119.

Parra, Estaban J., Rick A. Kittles, and Mark D. Shriver. 2004. "Implications of

Correlations between Skin Color and Genetic Ancestry for Biomedical Research." *Nature Genetics* 36 (11 suppl.): S54–S60.

Pfeifer, Gerd P., Young-Hyun You, and Ahmad Besaratinia. 2005. "Mutations Induced by Ultraviolet Light." *Mutation Research* 571 (1–2): 19–31.

Polhemus, Ted. 2004. *Hot Bodies, Cool Styles: New Techniques in Self-Adornment.* Photographs by UZi PART B. London: Thames and Hudson.

Race, Ethnicity, and Genetics Working Group. 2005. "The Use of Racial, Ethnic, and Ancestral Categories in Human Genetics Research." *American Journal of Human Genetics* 77 (4): 519–532.

Ragas, Meg C., and Karen Kozlowski. 1998. *Read My Lips: A Cultural History of Lipstick.* San Francisco: Chronicle Books.

Rana, Brinda K., David Hewett-Emmett, Li Jin, Benny H.-J. Chang, Namykhishing Sambuughin, Marie Lin, Scott Watkins, Michael Bamshad, Lynn B. Jorde, Michele Ramsay, Trefor Jenkins, and Wen-Hsiung Li. 1999. "High Polymorphism at the Human Melanocortin 1 Receptor Locus." *Genetics* 151 (4): 1547–1557.

Randle, Henry W. 1997. "Suntanning: Differences in Perceptions throughout History." *Mayo Clinic Proceedings* 72 (5): 461–466.

Ravage, Barbara. 2004. *Burn Unit: Saving Lives after the Flames.* Cambridge, Mass.: Da Capo Press.

Richardson, M. 2003. "Understanding the Structure and Function of the Skin." *Nursing Times* 99 (31): 46–48.

Robins, Ashley H. 1991. *Biological Perspectives on Human Pigmentation.* Vol. 7, *Cambridge Studies in Biological Anthropology,* edited by G. W. Lasker, C. G. N. Mascie-Taylor, and D. F. Roberts. Cambridge: Cambridge University Press.

Roddy, A. R., and J. D. Stosz. 1997. "Fingerprint Features: Statistical Analysis and System Performance Estimates." *Proceedings of the IEEE* 85 (9): 1390–1421.

Rogers, Alan R., David Iltis, and Stephen Wooding. 2003. "Genetic Variation at the MC1R Locus and the Time since Loss of Human Body Hair." *Current Anthropology* 45 (1): 105–108.

Rothschild, Lynn J. 1999. "The Influence of UV Radiation on Protistan Evolution." *Journal of Eukaryotic Microbiology* 46 (5): 548–555.

Ruff, Christopher B. 1991. "Climate and Body Shape in Hominid Evolution." *Journal of Human Evolution* 21 (2): 81–105.

Ruppenthal, Gerald C., G. L. Arling, Harry F. Harlow, Gene P. Sackett, and Stephen J. Suomi. 1976. "A 10–Year Perspective of Motherless-Mother Monkey Behavior." *Journal of Abnormal Psychology* 85 (4): 341–349.

Ruvolo, Maryellen. 1997. "Molecular Phylogeny of the Hominoids: Inferences from Multiple Independent DNA Sequence Data Sets." *Molecular Biology and Evolution* 14 (3): 248–265.

Sacks, Oliver. 1996. *An Anthropologist on Mars.* New York: Vintage Books.

Saikawa, Saito, Kimiko Hashimoto, Masaya Nakata, Masato Yoshihara, Kiyoshi Nagai, Motoyasu Ida, and Teruyuki Komiya. 2004. "The Red Sweat of the Hippopotamus." *Nature* 429 (6990): 363.

Sanchez-Vives, Maria V., and Mel Slater. 2005. "From Presence to Consciousness through Virtual Reality." *Nature Reviews: Neuroscience* 6 (4): 332–339.

Sapolsky, Robert M. 2004. *Why Zebras Don't Get Ulcers.* 3rd ed. New York: Owl Books.

———. 2005. "The Influence of Social Hierarchy on Primate Health." *Science* 308 (5722): 648–652.

Sarna, Tadeusz, and Harold M. Swartz. 1998. "The Physical Properties of Melanins." In *The Pigmentary System: Physiology and Pathophysiology,* edited by James J. Nordlund, Raymond E. Boissey, Vincent J. Hearing, Richard A. King, William Oetting, and Jean-Paul Ortonne. New York: Oxford University Press.

Sathian, Krishnankutty. 2005. "Visual Cortical Activity during Tactile Perception in the Sighted and the Visually Deprived." *Developmental Psychobiology* 46 (3): 279–286.

Saville, William J. V. 1926. *In Unknown New Guinea.* London: Seeley Service.

Saxe, Leonard. 1991. "Science and the CQT Polygraph: A Theoretical Critique." *Integrative Physiological and Behavioral Science* 26 (3): 223–231.

Schmidt, Karen L., and Jeffrey F. Cohn. 2001. "Human Facial Expressions as Adaptations: Evolutionary Questions in Facial Expression Research." *Yearbook of Physical Anthropology* 44:3–24.

Scott, M. Cathy, Itaru Suzuki, and Zalfa A. Abdel-Malek. 2002. "Regulation of the Human Melanocortin 1 Receptor Expression in Epidermal Melanocytes by Paracrine and Endocrine Factors and by Ultraviolet Radiation." *Pigment Cell Research* 15 (6): 433–439.

Scott, M. Cathy, Kazumasa Wakamatsu, Shosuke Ito, Ana Luisa Kadekaro, Nobuhiko Kobayashi, Joanna Groden, Renny Kavanagh, Takako Takakuwa, Victoria Virador, Vincent J. Hearing, and Zalfa A. Abdel-Malek. 2002. "Human Melanocortin 1 Receptor Variants, Receptor Function, and Melanocyte Response to UV Radiation." *Journal of Cell Science* 115 (11): 2349–2355.

Serre, David, André Langaney, Mario Chech, Maria Teschler-Nicola, Maja Paunovic, Philippe Mennecier, Michael Hofreiter, Göran Possnert, and Svante Pääbo. 2004. "No Evidence of Neandertal mtDNA Contribution to Early Modern Humans." *Public Library of Science: Biology* 2 (3): E57.

Shearn, Don, Erik Bergman, Katherine Hill, Andy Abel, and L. Hinds. 1990. "Facial Coloration and Temperature Responses in Blushing." *Psychophysiology* 27 (6): 687–693.

Sheridan, Robert L. 2003. "Burn Care: Results of Technical and Organizational Progress." *Journal of the American Medical Association* 290 (6): 719–722.

Silk, Joan B., Susan C. Alberts, and Jeanne Altmann. 2003. "Social Bonds of Female Baboons Enhance Infant Survival." *Science* 302 (5648): 1231–1234.

Sinha, Rajita, William R. Lovallo, and Oscar A. Parsons. 1992. "Cardiovascular Differentiation of Emotions." *Psychosomatic Medicine* 54 (4): 422–435.

Sinni-McKeehen, Barbara. 1995. "Health Effects and Regulation of Tanning Salons." *Dermatology Nursing* 7 (5): 307–312.

Smith, Rachel M., Eugene Healy, Shazia Siddiqui, Niamh Flanagan, Peter M. Steijlen, Inger Rosdahl, Jon P. Jacques, Sarah Rogers, Richard Turner, Ian J. Jackson, Mark A. Birch-Machin, and Jonathan L. Rees. 1998. "Melanocortin 1 Receptor Variants in an Irish Population." *Journal of Investigative Dermatology* 111 (1): 119–122.

Someya, Takao, Yusaku Kato, Tsuyoshi Sekitani, Shingo Iba, Yoshiaki Noguchi, Yousuke Murase, Hiroshi Kawaguchi, and Takayasu Sakurai. 2005. "From the Cover: Conformable, Flexible, Large-Area Networks of Pressure and Thermal Sensors with Organic Transistor Active Matrixes." *Proceedings of the National Academy of Sciences U.S.A.* 102 (35): 12321–12325.

Someya, Takao, and Takayasu Sakurai. 2003. "Integration of Organic Field-Effect Transistors and Rubbery Pressure Sensors for Artificial Skin Applications." *International Electron Devices Meeting '03 Technical Digest. IEEE International* 8.4.1–8.4.4.

Someya, Takao, Tsuyoshi Sekitani, Shingo Iba, Yusaku Kato, Hiroshi Kawaguchi, and Takayasu Sakurai. 2004. "A Large-Area, Flexible Pressure Sensor Matrix with Organic Field-Effect Transistors for Artificial Skin Applications." *Proceedings of the National Academy of Sciences U.S.A.* 101 (27): 9966–9970.

Spearman, R. I. C. 1977. "Keratins and Keratinization." In *Comparative Biology of Skin,* edited by R. I. C. Spearman. London: Academic Press.

Stanzl, Klaus, and Leonhard Zastrow. 1995. "Melanin: An Effective Photoprotectant against UV-A Rays." In *Melanin: Its Role in Human Photoprotection,* edited by Lisa Zeise, Miles R. Chedekel, and Thomas B. Fitzpatrick. Overland Park, KS: Valdenmar Publications.

Stringer, Chris. 2003. "Human Evolution: Out of Ethiopia." *Nature* 423 (6941): 692–695.

Sturm, Richard A. 2002. "Skin Colour and Skin Cancer—*MC1R,* the Genetic Link." *Melanoma Research* 12 (5): 405–416.

Sturm, Richard A., David L. Duffy, Neil F. Box, Wei Chen, Darren J. Smit, Darren L. Brown, Jennifer L. Stow, J. Helen Leonard, and Nicholas G. Martin. 2003. "The Role of Melanocortin-1 Receptor Polymorphism in Skin Cancer Risk Phenotypes." *Pigment Cell Research* 16 (3): 266–272.

Sturm, Richard A., Rohan D. Teasdale, and Neil F. Box. 2001. "Human Pigmentation Genes: Identification, Structure, and Consequences of Polymorphic Variation." *Gene* 277 (1–2): 49–62.

Suh, Jae Rin, A. Katherine Herbig, and Patrick J. Stover. 2001. "New Perspectives on Folate Catabolism." *Annual Review of Nutrition* 21:255–282.

Sulaimon, Shola S., and Barbara E. Kitchell. 2003. "The Biology of Melanocytes." *Veterinary Dermatology* 14 (2): 57–65.

Suomi, Stephen J. 1995. "Touch and the Immune System in Rhesus Monkeys." In *Touch in Early Development,* edited by T. M. Field. Mahwah, N.J.: Lawrence Erlbaum.

Taylor, Susan C. 2002. "Skin of Color: Biology, Structure, Function, and Implications for Dermatologic Disease." *Journal of the American Academy of Dermatology* 46 (2): S41–S62.

———. 2003. *Brown Skin: Dr. Susan Taylor's Prescription for Flawless Skin, Hair, and Nails.* New York: HarperCollins.

Thody, Anthony J., Elizabeth M. Higgins, Kazumasa Wakamatsu, Shosuke Ito, Susan A. Burchill, and Janet M. Marks. 1991. "Pheomelanin as well as Eumelanin Is Present in Human Epidermis." *Journal of Investigative Dermatology* 97 (2): 340–344.

Tsao, Hensin, and Arthur J. Sober. 2003. "Atypical Melanocytic Nevi." In *Fitzpatrick's Dermatology in General Medicine,* edited by Irwin M. Freedberg, Arthur Z. Eisen, Klaus Wolff, K. Frank Austen, Lowell A. Goldsmith, and Stephen I. Katz. 6th ed. New York: McGraw-Hill.

Underhill, Peter A., Peidong Shen, Alice A. Lin, Li Jin, Giuseppe Passarino, Wei H. Yang, Erin Kauffman, Batsheva Bonné-Tamir, Jaume Bertranpetit, Paolo Francalacci, Muntaser Ibrahim, Trefor Jenkins, Judith R. Kidd, S. Qasim Mehdi, Mark T. Seielstad, R. Spencer Wells, Alberto Piazza, Ronald W. Davis, Marcus W. Feldman, L. Luca Cavalli-Sforza, and Peter J. Oefner. 2000. "Y Chromosome Sequence Variation and the History of Human Populations." *Nature Genetics* 26 (3): 358–361.

Van Boven, Robert W., Roy H. Hamilton, Thomas Kauffman, Julian P. Keenan, and Alvaro Pascual-Leone. 2000. "Tactile Spatial Resolution in Blind Braille Readers." *Neurology* 54 (12): 2230–2236.

Vasold, Rudolf, Natascha Naarmann, Heidi Ulrich, Daniela Fischer, Burkhard König, Michael Landthaler, and Wolfgang Bäumler. 2004. "Tattoo Pigments Are Cleaved by Laser Light—The Chemical Analysis *in vitro* Provide Evidence for Hazardous Compounds." *Photochemistry and Photobiology* 80 (2): 185–190.

Vermeij, Geerat J. 1999. "The World According to the Hand: Observation, Art, and Learning through the Sense of Touch." *Journal of Hand Surgery* 24A:215–218.

von Luschan, Felix. 1897. *Beitrage zur volkerkunde der deutschen Schutzgebiete.* Berlin.

Wadman, Meredith. 2005. "Scar Prevention: The Healing Touch." *Nature* 436 (7054): 1079–1080.

Wagner, Jennifer K., Esteban J. Parra, Heather L. Norton, Celina Jovel, and Mark D. Shriver. 2002. "Skin Responses to Ultraviolet Radiation: Effects of Constitutive Pigmentation, Sex, and Ancestry." *Pigment Cell Research* 15 (5): 385–390.

Walker, Alan, and Richard E. Leakey, eds. 1993. *Nariokotome* Homo erectus *Skeleton.* Cambridge, Mass.: Harvard University Press.

Walsberg, Glenn E. 1988. "Consequences of Skin Color and Fur Properties for Solar Heat Gain and Ultraviolet Irradiance in Two Mammals." *Journal of Comparative Physiology B* 158 (2): 213–221.

Walter, Philippe, Pauline Martinetto, Georges Tsoucaris, Rene Bréniaux, M.A. Lefebvre, G. Richard, J. Talabot, and Eric Dooryhée. 1999. "Making Make-Up in Ancient Egypt." *Nature* 397 (6719): 483–484.

Wassermann, Hercules P. 1965. "Human Pigmentation and Environmental Adaptation." *Archives of Environmental Health* 11 (5): 691–694.

———. 1974. *Ethnic Pigmentation.* New York: American Elsevier.

Webb, Ann R., and Michael F. Holick. 1988. "The Role of Sunlight in the Cutaneous Production of Vitamin D_3." *Annual Review of Nutrition* 8:375–399.

Webb, Ann R., L. Kline, and Michael F. Holick. 1988. "Influence of Season and Latitude on the Cutaneous Synthesis of Vitamin D_3: Exposure to Winter Sunlight in Boston and Edmonton Will Not Promote Vitamin D_3 Synthesis in Human Skin." *Journal of Clinical Endocrinology and Metabolism* 67 (2): 373–378.

Weze, Clare, Helen L. Leathard, John Grange, Peter Tiplady, and Gretchen Stevens. 2005. "Evaluation of Healing by Gentle Touch." *Public Health* 119 (1): 3–10.

Wharton, Brian, and Nick Bishop. 2003. "Rickets." *The Lancet* 362 (9393): 1389–1400.

Wheeler, Peter E. 1984. "The Evolution of Bipedality and Loss of Functional Body Hair in Hominids." *Journal of Human Evolution* 13:91–98.

———. 1985. "The Loss of Functional Body Hair in Man: The Influence of Thermal Environment, Body Form, and Bipedality." *Journal of Human Evolution* 14:23–28.

———. 1988. "Stand Tall and Stay Cool." *New Scientist* 118:62–65.

———. 1991a. "The Influence of Bipedalism on the Energy and Water Budgets of Early Hominids." *Journal of Human Evolution* 21 (2): 117–136.

———. 1991b. "The Thermoregulatory Advantages of Hominid Bipedalism in Open Equatorial Environments: The Contribution of Increased Convective Heat Loss and Cutaneous Evaporative Cooling." *Journal of Human Evolution* 21 (2): 107–115.

White, Tim D., Berhane Asfaw, David DeGusta, Henry Gilbert, Gary D. Richards,

Gen Suwa, and F. Clark Howell. 2003. "Pleistocene *Homo sapiens* from Middle Awash, Ethiopia." *Nature* 423 (6941): 742–747.

Whitear, Mary. 1977. "A Functional Comparison between the Epidermis of Fish and of Amphibians." In *Comparative Biology of Skin*, edited by R.I.C. Spearman. London: Academic Press.

Widdowson, Elsie M. 1951. "Mental Contentment and Physical Growth." *The Lancet* 1 (24): 1316–1318.

Wilkin, Jonathan K. 1988. "Why Is Flushing Limited to a Mostly Facial Cutaneous Distribution?" *Journal of the American Academy of Dermatology* 19 (2, pt. 1): 309–313.

Wood, Fiona. 2003. "Clinical Potential of Autologous Epithelial Suspension." *Wounds* 15 (1): 16–22.

Wu, Ping, Lianhai Hou, Maksim Plikus, Michael Hughes, Jeffrey Scehnet, Sanong Suksaweang, Randall B. Widelitz, Ting-Xin Jiang, and Cheng-Ming Chuong. 2004. "*Evo-Devo* of Amniote Integuments and Appendages." *International Journal of Developmental Biology* 48 (2–3):249–270.

Yankee, William J. 1995. "The Current Status of Research in Forensic Psychophysiology and Its Application in the Psychophysiological Detection of Deception." *Journal of Forensic Science* 40 (1): 63–68.

Yee, Ying K., Subba R. Chintalacharuvu, Jianfen Lu, and Sunil Nagpal. 2005. "Vitamin D Receptor Modulators for Inflammation and Cancer." *Mini Reviews in Medicinal Chemistry* 5 (8): 761–778.

Young, Antony R. 1997. "Chromophores in Human Skin." *Physics in Medicine and Biology* 42:789–802.

Young, Antony R., and John M. Sheehan. 2001. "UV-Induced Pigmentation in Human Skin." In *Sun Protection in Man*, edited by Paolo U. Giacomoni. Amsterdam: Elsevier.

Yue, Zhicao, Ting-Xin Jiang, Randall Bruce Widelitz, and Cheng-Ming Chuong. 2005. "Mapping Stem Cell Activities in the Feather Follicle." *Nature* 438 (7070): 1026–1029.

Zenz, Rainier, Robert Eferl, Lukas Kenner, Lore Florin, Lars Hummerich, Denis Mehic, Harald Scheuch, Peter Angel, Erwin Tschachler, and Erwin F. Wagner.

2005. "Psoriasis-like Skin Disease and Arthritis Caused by Inducible Epidermal Deletion of Jun Proteins." *Nature* 437 (7057): 369–375.

Zihlman, Adrienne L., and Bruce A. Cohn. 1988. "The Adaptive Response of Human Skin to the Savanna." *Human Evolution* 3 (5): 397–409.

Zouboulis, Christos C. 2000. "Human Skin: An Independent Peripheral Endocrine Organ." *Hormone Research* 54 (5–6): 230–242.

INDEX

DESIGNER
Nicole Hayward

TEXT
9.75/15 Scala

DISPLAY
Interstate

COMPOSITOR
Integrated Composition Systems

INDEXER
Barbara Roos

ILLUSTRATOR
Jennifer Kane

PRINTER AND BINDER
Friesens Corporation